Thomas M. Goodeve

Principles of Mechanics

Second Edition

Thomas M. Goodeve

Principles of Mechanics
Second Edition

ISBN/EAN: 9783337311544

Printed in Europe, USA, Canada, Australia, Japan

Cover: Foto ©berggeist007 / pixelio.de

More available books at **www.hansebooks.com**

PRINCIPLES

OF

MECHANICS.

BY

T. M. GOODEVE, M.A.

Barrister-at-Law :

Lecturer on Applied Mechanics at the Royal School of Mines.

SECOND EDITION.

LONDON:

LONGMANS, GREEN, AND CO.

1876.

PREFACE.

THIS BOOK contains an outline of one part of the course of lectures on Applied Mechanics given by the writer at the Royal School of Mines. It is impossible, within the limits of this volume, to present a finished sketch of the principles of Mechanics. Still something may be done. An endeavour may be made to present a comprehensive view of the science, to point out the necessity of continually referring to practice and experience, and above all to show that the relation of the theory of heat to mechanics should be approached by the student, in his earliest inquiries, with the same careful thought with which he will surely regard it when his knowledge and his powers have become extended and strengthened.

T. M. G

TEMPLE : *March* 1874.

CONTENTS.

—◦—

INTRODUCTION.

CHAPTER I.

THE PARALLELOGRAM OF FORCES AND THE LEVER.

CHAPTER II.

ON WORK AND FRICTION.

CHAPTER III.

ON THE CENTRE OF GRAVITY.

CHAPTER IV.

ON SOME OF THE MECHANICAL POWERS.

CHAPTER V

ON THE EQUILIBRIUM AND PRESSURE OF FLUIDS.

CHAPTER VI.

ON THE EQUILIBRIUM AND PRESSURE OF GASES

CHAPTER VII.

ON PUMPS.

CHAPTER VIII.

ON THE HYDRAULIC PRESS AND HYDRAULIC CRANES.

CHAPTER IX.

ON MOTION IN ONE PLANE.

CHAPTER X.

ON CIRCULAR MOTION.

CHAPTER XI.

ON GIRDER BEAMS AND BRIDGES, THE STRENGTH OF TUBES, AND THE CATENARY.

CHAPTER XII.

ON SOME MECHANICAL INVENTIONS.

MECHANICS.

No effectual progress can be made in the study of Mechanics by those who fail to understand the principles which have guided scientific men in an endeavour to measure with precision certain varied and mysterious actions by which we are surrounded, and to which, though it is almost hopeless to attempt to define the exact meaning of the term, we have given the name of Force.

Whatever idea may be formed of the nature of force, some mode of measuring it must be agreed upon, and the first portion of this chapter will be devoted to an enquiry into the reasons which have led to the adoption of two different units for the measurement of force.

For the present purpose it will suffice to regard *force as any cause which moves, or tends to move, any portion of matter.*

In order to conceive the existence of a force we must also conceive that there is something upon which it can act, and accordingly we define *matter* as being anything which may be perceived by the senses, and which *can be acted upon by, or can exert, force.*

Our ideas of matter and force coexist together. We shall soon learn to connect the idea of force with that of matter in motion, and shall further regard matter as the depository of force. In this way of treating the subject it is customary to speak of *powerful* matter or of *powerless* matter. A moving weight or a raised weight is powerful as distinguished from the same weight after it has come to rest, or after it has fallen—that is, after the power has left it. So also, and for a more subtle reason, a piece of coal is powerful, and the gases, into which it burns, are powerless. We can only comprehend such statements by investigating the primary laws which govern the action of forces upon matter, and in doing so we purpose to follow the course laid down by Newton, and to enunciate the principles or laws of motion upon which this science of mechanics is so firmly founded. It will be necessary, in the first instance, to prepare the way by a few preliminary remarks and definitions.

THE REPRESENTATION OF FORCES.

Forces are represented by straight lines.—In this way many mechanical problems are solved by the aid of geometry.

There are three things to be known before a force can be said to be fully determined :

1. The point upon which it acts.
2. Its direction, or the line in which it acts.
3. Its magnitude.

The point upon which a force is supposed to act must be a material point, or that extremely minute portion of matter which has been named a molecule. It is not a geometrical point, but may almost be regarded as such.

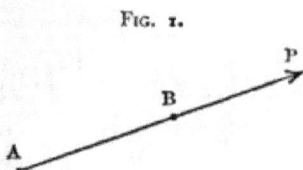

FIG. 1.

Conceive that a force of P units, called a force P, acts upon a point A ; the point A will begin to move in some determinate direction as A P, and this direction is the *line of action* of the force P.

Hence we represent the point upon which the force P acts by the point A in the line A P, and we regard the direction of the straight line A P as the direction of the force.

It only remains to represent the magnitude of the force; this can be **done by** taking A B equal to as many units of length as **there are units of** magnitude in the force, and the line A B **will then** completely represent in magnitude and direction the force P **acting at A.**

Definition.—*Two forces which act* **in** *opposite directions in the same straight line upon* **a material free point, and do** *not* **move** *it, are said to be equal.*

Forces are also equal when they **act in** opposite directions at *separated* points in a straight line, and balance each other.

The points must be rigidly connected, and we proceed to state a principle called *the principle of the transmission* of *force through* **a rigid body,** which enables us to attach an **ex**-tended meaning to **the phrase** 'line of action of a force.'

If a force **act** *upon a* **rigid** *body* (meaning thereby a body which **is not** susceptible of change of form), *it is transmitted* **by the rigid** *body along its line of direction, and may be supposed to act at* **any** *point of that line.*

This is equally true whether the line of action of the force is contained within the body or not.

Admitting this principle of the transmission of force, which we derive from experiment and observation, it is evident that we may add or subtract forces whose lines **of** direction coincide, just as **we can** add or subtract any **other** magnitudes.

Also since **forces are** completely represented by straight lines, we can conceive that forces may have any ratio to each **other in** respect of magnitude.

THE GRAVITATION OF MATTER.

The *theory of* **universal** *gravitation,* as it **is** termed, **was** established by Newton. No truth in science **can be deemed** to rest upon a more **sure** foundation. According to this theory, every particle of matter is **endowed with a power of** attracting or drawing towards it **every other particle,** wherever situated ; and further, **the attractive power of matter**

at different distances diminishes in the same proportion as the square of the number expressing the distance increases.

The force of attraction or gravitation is associated inseparably with matter of every kind : no art, or device, or power can remove, change, or diminish it in any degree. It is the one unalterable but unexplained power attaching to material substances, and is described as universal because there are strong grounds for asserting that its influence is felt far beyond the limits of our planetary system.

The amount of this force which is inherent in small bodies such as a stone or a ball of lead is so trifling that it cannot be recognised, but, like many other minute actions, it becomes very apparent by accumulation, and the attraction of the matter constituting the earth evidences itself by causing every substance to press towards the centre with a constant and never-ceasing power. Hence all substances within our observation have the quality of *weight*, or exert a pressure upon their supports, the weight of the body being this pressure or tendency to move downwards by reason of the attracting force of the whole globe of the earth.

UNIFORM AND VARIABLE FORCES.

It will soon become important to examine whether the attracting force of the earth is uniform or variable, and we may now pause to consider what is meant by a *uniform force.* The phrase is evidently intended to apply to a force which in no respect changes its intensity or power of action during the time that we are subjecting it to investigation.

Many forces treated of in mechanics are *variable* forces : thus the force of recoil exerted by a spiral steel spring after it has been stretched is a variable force. The power necessary to stretch an elastic spring increases with the ›
amount of the elongation, and the recoil varies in intensity in like manner. The attraction exerted by the pole of a magnet on a piece of iron depends on the distance of the

magnet from the iron, and varies accordingly; this attraction is therefore not a constant force.

On the other hand, it may be **proved** that the attraction of the earth is a constant or uniform force for bodies at the same part of its surface, and the attractive power of the earth is often quoted as an example of a constant force everywhere on the globe, although in truth it changes a little in different latitudes. But the gravitation towards the centre of the earth is essentially a variable force when we come to deal with distances beyond its surface : thus the attraction of the earth on a body 20,000 miles distant from its centre is sensibly greater than if it were ten times farther off, and is, in fact, one hundred times as great.

The truth of this observation was first established by Newton, who proved that the attraction of the earth held the moon in her orbit by a force which varied from that upon the surface, according to the law to which reference has been made. Since that time, the theory of the motion of the moon, the calculations of the moon's place as made for the purposes of navigation, and the theory of the motion of the planets, have all proceeded upon the supposition that the attractive force of any large mass of matter, such as the earth, is sensibly variable, and changes with the distance.

An extended knowledge of the calculus is essential for the student who desires to comprehend the action of variable forces. Something may be done by simple geometry; and in order to account for the observed motions of the planets round the sun, Newton employed geometrical reasoning of extraordinary beauty and elegance. But the geometry of Newton has been compared to the bow of Ulysses, which he alone could handle with effect, and is not adapted for discussion in an elementary work.

It may, therefore, simplify matters if we point out that every force referred to in this chapter or subsequently will be deemed to be a *constant* or *uniform force* unless the contrary is distinctly stated. This proviso is very necessary, as the reader would convert many propositions into absolute non-sense if he were to extend them to the action of variable forces.

THE GRAVITATION MEASURE OF FORCE.

A standard pound avoirdupois is a certain piece of platinum which is deposited in the office of the Exchequer, and which serves as the standard of weight in this country.

There are two things that we notice with reference to this standard pound :

1. It contains an invariable quantity of matter.

2. It has the quality of weight ; that is, it is pulled downwards by the attraction of the earth.

If the pound were hung upon a spring balance, it would stretch the spring by a definite amount, and might replace muscular force.

Regarding this piece of platinum from two different points of view, we might use it either to measure the quantities of matter in different bodies, and thus assign numerical values representing those quantities, or we might employ it as a standard for estimating the magnitude of forces, and might institute a comparison between any given force and the pull of the earth upon the standard piece of platinum.

The reader should now be made aware that by the *mass* of a body we understand the quantity of matter in it : this term is never used in any other sense ; and it is at once evident that it would be a very natural and convenient course to select this piece of platinum for the *unit of mass*, and to estimate the masses of all bodies by the process of weighing them against this standard weight or others derived from it.

A unit must be an invariable quantity, and this is clearly an unchangeable mass of matter : if it were possible to transport it to another planet, its mass would not be altered thereby ; it remains the same when carried from one place to another, and possesses the one great quality which we require in a unit, viz. that of invariability.

But the science of mechanics has grown up insensibly from the earliest ages in the history of the world, and the pound weight was made a standard for the wants of civilised life long before there was any thought of such a fact as that the attraction of the earth really caused a piece of metal to have weight; thus the *gravitation measure of force*, or the estimation of forces by the weights they will support, came into general use, not for any scientific reason, but because it afforded the most ready and simple method of estimating the forces applied for any given purpose.

In truth, the pound weight has become appropriated, as it were, for the measurement of force rather than for that of mass, and is the recognised standard of reference, not only for force, but also for the work done by force : such expressions as a force of ten pounds, a pressure of steam equal to fifty pounds on the inch, and the like, being of every-day occurrence.

The piece of platinum called a pound is clearly a correct measure of mass, but is it also a correct measure of force? In other words, is the weight of this standard piece of platinum absolutely invariable? If it be invariable, it may rightly serve as a unit of force ; but if it be liable to change, it fails in the primary requisite. Everyone knows that the earth is not a perfect sphere, that it is flattened at the poles, and that it rotates once in twenty-four hours. From these facts it is possible to arrive at the conclusion that the pull of the earth upon our standard piece of platinum is a little greater in London than it is at the Equator. The difference is very trifling, and can only be discovered by experiments of extreme delicacy ; but it exists notwithstanding : and there is abundant evidence that our standard pound has not the essential quality of a unit when used as a measure of weight for the earth generally. We have stated that the variation is scarcely appreciable, and Sir John Herschel informs us that the pull of the earth or the force of gravity in London is to that at the Equator in the proportion of the numbers 100,315 to 100,000. There are 7,000 grains in a pound avoirdupois, and therefore the downward tendency of a pound weight at the Equator would be less than that in London by about twenty-two grains, that is to say, 6,978 parts of the 7,000 of which our platinum standard pound consists, would be supported in London by the same extraneous force as the whole 7,000 grains at the equator.

A variation so small as this is of no consequence in practice; it is only where extraordinary accuracy is required that the difficulty would press upon us: nevertheless, a science ought to be erected upon a solid foundation, and the science of mechanics, which is to be applied to investigate the action of forces of the most varied kind, such as the pressure of the air or the attraction of an electrified ball, should be based upon measurements which are not vitiated by errors, however minute, and of which our reason will therefore approve.

THE RELATION BETWEEN FORCE AND MOTION.

If the pound weight were abandoned as the measure of force, what other measure could be suggested?

To answer this enquiry we must return to our piece of platinum and see how it behaves under the action of force. If allowed to fall, the pull of the earth would immediately cause it to move with increasing rapidity; if it were placed on a smooth table, a sustained push by the hand would do the same thing: it is evident that the effect of force is to produce motion, and that the quantity of motion produced, if we could agree upon some method of estimating it, may be taken as a measure of force.

But a vague notion, such as that here presented, has no scientific value, and we pass on, therefore, to a critical examination of the fundamental laws of the action of force in producing motion, which must be discussed at the first stage of our progress, if we are to begin this study by learning how to measure force.

Some definite ideas on the constitution of matter may be suggested as introductory to the enquiry.

Every substance, a piece of glass for example, is to be regarded as the collection of a multitude of small parts called molecules. Each molecule consists of a finite quantity of matter, and may itself be composed of other portions of matter which are held together by chemical bonds of union. Thus the oxide of lead necessary for the manufacture of flint glass consists of compound molecules whose separate parts, viz. oxygen and lead, are held together by chemical force.

Again, the molecules of all bodies within our observation are not at rest, but are in a state of continual and never-ceasing agitation ; this agitation grows more intense when the body is heated, and dies away as the body becomes cooler, but it is never entirely quenched or put an end to by any degree of cold which we can produce. The agitation or vibration to which we have referred cannot be detected by any observation, and the most powerful microscopes fail to reveal to us any trace of this movement. All that we can recognise in matter is perfect repose, without even a suggestion of visible motion : nevertheless a firm conviction of the truth of the hypothesis above stated has taken hold of men's minds ; the modern theory of heat is based upon it, and the evidence in support of that theory appears to be as reliable as that upon which we ground our belief in the doctrine of universal gravitation.

It is believed that sensible heat is motion ; and further, that there can be no exhibition of force upon matter without the expenditure of heat, that is, without condensing and accumulating upon some isolated body the minute agitations which reside in a multitude of these molecules, the tremulous motion of which will be reduced by a quantity exactly equal to that imparted to the mass upon which the force is acting.

It cannot, therefore, be doubted that there is abundant

reason for entering upon the study of mechanics by an enquiry into the action of force upon matter.

THE MEASUREMENT OF VELOCITY, WHETHER LINEAR OR ANGULAR.

The term *velocity* is employed in a technical sense, and expresses the degree of swiftness or rapidity with which a body is moving.

A body can move in two different ways :

1. It may have a *motion of translation*, such as that of a stone thrown up into the air ; in that case every point in the body will move with the same velocity, and every straight line in the body will remain parallel to itself. Thus the disc A has simply a motion of translation in passing, as in the diagram, from A to B. The arrow does not change its direction.

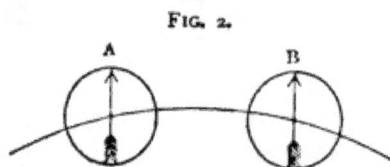

FIG. 2.

2. A body may have a *motion of rotation* ; that is, it may spin like a top. In such a case the lines in the body will change their direction, and every point in the body will describe a circle whose plane is perpendicular to the axis or line about which the body is rotating.

These two motions frequently exist together, as seen in fig. 3, where the disc rotates as it moves from A to B. The moon has a motion of translation and also one of rotation ; in virtue of the former, it describes an orbit round the earth, wh reas the latter causes it to turn upon its axis so as always

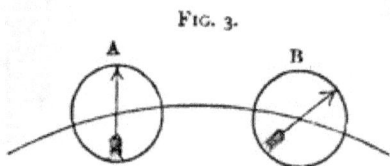

FIG. 3.

to present the same face for our observation. A rifle bullet has the two movements : the bullet spins on its axis while describing a somewhat curved path in the air.

On account of the possibility of these two motions we have to distinguish *velocity* as being *linear* or *angular* Linear velocity refers to a motion of translation, and angular velocity to one of rotation.

In measuring velocities it is usual to adopt a *second* and a *foot* as the units of time and length. The number of feet which a moving body passes over in a given time is called the *space* described in that time.

Velocity is said to be uniform when equal spaces are described in equal times by the moving body.

The *linear velocity* of a body *when uniform* is measured by the number of feet described in one second. When we speak of a body having a velocity of thirty feet, we mean thereby that thirty is the number of feet which would be described if the body were to move uniformly for one second with the velocity which it has at the instant considered.

Thus if v be the velocity of a body moving uniformly,

s the space described in t seconds,

the letter v represents the number of feet described in one second, and the motion is uniform : therefore $2v$ represents the number of feet described in two seconds, and tv the number described in t seconds.

Hence $$s = tv.$$

Ex. Find the space described in three seconds by a body moving uniformly with a velocity of twenty miles per hour.

Since the body describes 20×5280 feet in 60×60 seconds, it describes $\dfrac{3 \times 20 \times 5280}{60 \times 60}$ in 3 seconds.

$$\therefore \text{ space required} = 88 \text{ feet.}$$

It is apparent that a velocity of one mile per hour is equivalent to a velocity of 88 feet per minute, or of $\dfrac{88}{60} = 1 \cdot 47$ feet per second.

Since $\dfrac{88}{60} = \dfrac{22}{15} = 1 + \dfrac{1}{2 + \frac{1}{7}}$,

we may approximate by writing $1 + \frac{1}{2}$ or $\frac{3}{2}$ for $\dfrac{22}{15}$, and thus it is common to estimate a velocity of x miles per hour as equivalent to a velocity of $\frac{3x}{2}$ feet per second, the accurate value being $\frac{22x}{15}$

The next consideration is the measurement of angular velocity.

It has been stated that every point in a body set in rotation about an axis will describe a circle lying in a plane perpendicular to the axis. Let A P B represent the circle described by any point P in a rotating body ; c the centre of the circle, that is, the point of intersection of its plane with the axis of rotation. The *angular velocity* of the body *when uniform* is measured by the angle described in one second by a line c P which revolves round c in a plane perpendicular to the axis of rotation.

FIG 4.

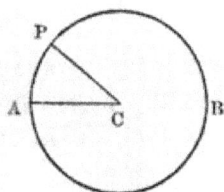

Hence if P moves from A to P in one second, the angular velocity of the body is the angle P C A.

This angle is not expressed in degrees, but in circular measure, that is by the ratio of A P to A C : it will be seen that this is a constant ratio for every point in a body rotating uniformly, and it has been customary to designate it by the Greek letter ω (*omega*).

Hence
$$\omega = \frac{A\ P}{A\ C}.$$

Let r be the radius of the circle A P B.

v the linear velocity of the point P.

then
$$v = A\ P, \quad \omega = \frac{A\ P}{C\ P} = \frac{v}{r}.$$

$$\therefore v = \omega\ r.$$

Which is the equation connecting the linear velocity of the point P with the angular velocity of the body. It follows that when a body is rotating uniformly, the linear velocity of any point in it increases directly as the distance from the axis of rotation.

Ex. 1. Suppose a wheel 12 feet in diameter to make 30 revolutions in one minute, find its **angular** velocity, and the linear velocity of a point in the rim.

The **wheel makes $\frac{1}{2}$ a** revolution **in one second, or** describes **an angle of 180 degrees in** one second. This **angle, expressed in circular measure, is equal to the ratio** $\dfrac{\frac{1}{2} \text{ circumference}}{\text{radius}}$.

Hence $\omega = \dfrac{\frac{1}{2} \text{ circumference}}{\text{radius}} = \pi = 3\cdot1416.$

The linear velocity of a point in the rim is therefore equal to $\pi \times 6$ feet $= 18\cdot8496$ feet.

Ex. 2. Suppose a straight rod to be set in rotation about an axis through its centre and perpendicular to its length, with such a velocity that the rod describes an angle of $534°$ in one second. Find the angular velocity of the rod, and the linear velocity of a point at a distance of ten inches from the axis.

Since $\qquad\qquad 534° = 360° + 174°$

we have $\qquad \omega = 2\pi + \dfrac{174}{180} \pi = \dfrac{534}{180} \pi = \dfrac{534}{180} . 3\cdot1416$

$$\therefore \ \omega = 9\cdot32008,$$

and the linear velocity required $= \omega \times \dfrac{10}{12}$ feet $= 7\cdot767$ feet.

THE FIRST LAW OF MOTION.

The final step in our preliminary enquiries will conduct us to an examination of the precise laws which govern the action of force upon matter. These laws have been stated by Newton with a clearness and precision which cannot be improved, and the student will do well to endeavour to appreciate the exact meaning of every word in the statement. It is only by the most careful attention to the meaning of each word and clause that the mind will be enabled to grasp the scientific truths involved in these brief sentences.

Newton's first law of motion is the following :

First **Law.**—*Every body continues in its state of rest or of uniform motion in a **straight line**, except in so far as it may be compelled by impressed forces to change that state.*

This law is intended to assert the *inertia* of matter, or

that quality inherent to matter whereby it has no power in itself to change its own state of rest or motion.

It will be readily understood that absolute rest nowhere exists in nature : the word *rest* may be understood in a relative sense.

Again, an *impressed force* is merely another phrase for a force : all *external forces* which act upon bodies are impressed forces.

The conclusion that matter has no spontaneous power of moving itself is quite irresistible, and is in accordance with the experience of daily life ; but the idea of permanence in motion, or the conception that all moving bodies, at every instant, are exerting a never-ceasing tendency to persist in one simple movement in a straight line with a uniform velocity, is not grasped without some difficulty. The elements of decay exist in all visible things, and we might possibly persuade ourselves that there is some principle of decay or diminution inherent in the nature of motion. If we did so, without doubt we should commit a grave error ; and it will be found that all experimental facts, and all reasoning upon observed facts, point to the opposite conclusion. The limits of this treatise permit only the statement of the law, and it must suffice to call attention to certain observations which support our belief in its truth.

There can be no question that all the movements which we perceive in bodies around us tend to diminish, and eventually to become extinct. In order to maintain bodies in motion we must continually make fresh efforts to replenish the waste ; nevertheless we do not refer this gradual subsidence of movement to any cause inherent in the body, but rather to the effect of retarding forces, such as friction or the resistance of the air, whose action in destroying the motion can be readily appreciated. In whatever degree we remove these retarding forces, we do but render more complete the assertion of this principle of the permanence of motion.

Thus the principle applies in such examples as the following :—

If a carriage be in steady motion, the occupants are very much in the same condition as if they were sitting in a room at rest ; but let the carriage be suddenly stopped, their motion will continue, and they will be thrown violently forward. Note also the difficulty of stopping a railway train at a station, or in pulling up a horse at full gallop, or in turning a corner when running, or in moving aside when skating rapidly. Why is it that a circus rider springs upwards, and not forwards, when leaping through a hoop—that a stone flies onward from a sling, that a pendulum does not stop suddenly in the middle of its swing, or that a light hammer will strike a heavy blow ? There is no limit to the number of such observations.

This law may be extended beyond the limits which appear to be assigned to it by Newton, and will be equally true when we apply its principle to the case of a body rotating about an axis. Such a body will continue in its state of uniform rotation except so far as it may be compelled by impressed forces to change that state.

Experiments on the spinning of a top would soon lead us to trace the subsidence of its motion to the friction of the pivot, and the action of the air ; but without pausing to discuss the obvious effect of these retarding forces, we may with advantage contemplate that wonderful example of the permanence of rotation which is afforded by the earth itself.

An experiment with a pendulum, suggested by M. Foucault, and founded upon this law of motion, has been employed to demonstrate the fact that the earth really does rotate upon its axis. In making this experiment we see the plane of swing of the pendulum slowly twisting round with a motion which can be caused by nothing else than a movement of a similar kind in the earth itself.

There are no disbelievers in the earth's rotation ; but the question now will be, is that rotation sensibly constant under the known circumstances of motion in that medium, rarer than all others, upon which we have conferred the name of empty space, and where the causes of retardation which we observe in a rotating wheel or top are completely abolished ?

Although we may readily allow that there exist no sensible resistances external to our planet, yet there are other agencies to be regarded. It appears certain that the friction caused by tidal action must exert some influence in diminishing the earth's rotation, and it is manifest that we have further to take into account the gradual cooling and shrinking of the entire mass.

One thing is well established, namely, that the change is so minute as to leave it doubtful whether any retardation has been completely ascertained. In referring to this subject it has been the practice to instance the calculations of Poisson, the mathematician, who assumed that the length of a day had diminished by one ten-millionth part since the period of the most ancient recorded eclipse, 720 B.C. Poisson estimated that the earth and moon were so placed that it would have been impossible for the moon to dip into the earth's shadow at the stated epoch if any such diminution had really occurred.*

But in recent years we have the more reliable investigation of Mr. Adams, the astronomer, who has pointed out that the theoretical value adopted for the so-called acceleration of the moon's motion was erroneous, and that the conclusion of Poisson was not well founded. Mr. Adams estimates that the earth would in a hundred years get twenty-two seconds behind a perfect clock rated at the beginning of the century.

Twenty-two seconds in arrear after all the revolutions made in a hundred years is a near approach to perfect uniformity, and may well have been mistaken for it. There is therefore nothing in this estimation which can affect our belief in the first law of motion. Hereafter when we connect the motion of masses with work done, and are made aware

* See *Grant's History of Physical Astronomy*, First Edition, p. 161.

of the amount of force necessary to interfere with the movement of so vast a body as the earth, we shall better comprehend the invariability of its rotation.

Not only does the earth rotate on its axis with undeviating uniformity, but the position of that axis is maintained by the rotation at an angle to the plane in which the earth sweeps round the sun, which is practically constant. The examination of this subject demands very extended knowledge, but we may refer to an illustration which exhibits the permanence due to rotation.

Let A B be the axis of a small heavy disc which can be set spinning, and which is supported on the ring A E B; this ring is pivoted so as to turn freely on the axis E *e*, and a stem, which fits loosely in the hollow tube D, is attached to E C *e*.

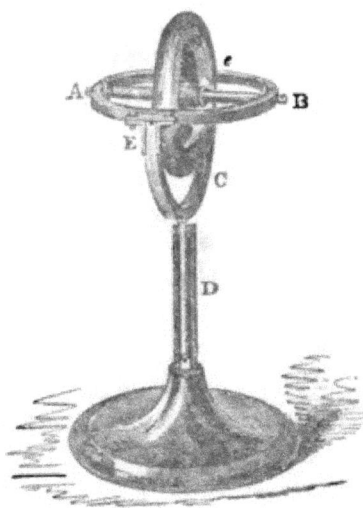

FIG. 5.

If the disc be at rest, the slightest tap at E will cause the apparatus to turn round, for the stem fits quite loosely, and there is nothing to prevent this motion.

Set now the disc in rapid rotation, and tap the ring as before; it will be found to be immovable, it will oppose a determined resistance, and the stem will appear to be locked in the hollow tube. Something, of course, must move, and the effect of the blow is to cause the ring A E B *e* to turn upon the axis E *e*.

The rotation of the disc produces this singular result; it is apparent that each portion of the disc is describing a circle in a plane perpendicular to its axis: the whole mass is therefore travelling round in parallel planes. If

C

the axis were to change its position, the motion would take place in a new set of planes inclined to the former. This change can only arise from the action of force, and the resistance which is felt is a direct consequence of the inertia of matter. If a person were to hold the stand in his hand and to walk round in a circle, the plane of the disc would appear to turn automatically in its support : this is another method of trying the experiment.

THE SECOND LAW OF MOTION.

The first law of motion in effect tells us that it is only extraneous force which can generate or destroy motion, or which can produce a change of motion. The definition of force will be framed in accordance with this law, and we pass on to examine the exact relation between force and the motion which it produces.

Second Law. *Change of motion is proportional to the impressed force, and takes place in the direction of the straight line in which the force is impressed.*

Hitherto it has been sufficient to speak of the motion of bodies, without regarding this motion as a measurable quantity which can be compared with other magnitudes such as forces. It is necessary now to attach some very definite ideas to the words 'change of motion.' And first we remark that the idea of motion commonly accepted, as meaning simply velocity, is not that which is here conveyed. Two bodies, of whatever weights, when moving with the same velocity, are often spoken of as having the same motion, whereas the quantity of motion existing in each may be very different. If a light and heavy body are moving with the same velocity, it will require a much greater exertion of force to stop the heavier body.

The term '*motion*' must be understood as signifying 'the *quantity of motion*,' or as it is otherwise named, '*momentum*, a word used in a scientific sense, and often strangely mis-applied. We proceed to interpret this expression, and shall

endeavour to affix to it the meaning which it must necessarily have if the law be true.

Conceive a pound weight to be at rest, and that an extraneous force acts upon it; the change of motion will now be
the whole motion impressed upon the weight. If the force
be doubled, the quantity of motion will also be doubled.
But the pound weight does not alter, and therefore we can
only change the quantity of motion by changing the velocity;
hence we infer that *the quantity of motion varies as the
velocity generated when the weight moved remains constant.*
Again, if a force act upon a pound weight for a given time,
the weight will acquire a certain velocity, and have a certain
quantity of motion impressed upon it. If the same force
act upon two pounds weight, one half of the force will be
felt upon each pound, and either pound weight will move
with half its previous velocity. But according to this law, a
second force, equal to the first, will impress the same
quantity of motion upon the now moving mass of two
pounds, from which we infer that when the double weight
moves with the same velocity as the single one, the extraneous force, and therefore also the quantity of motion, is
doubled. In other words, *when the velocity remains constant
the quantity of motion varies as the weight moved.*

Now there is a well-known principle in algebra and arith-
metic, which is, in fact, the double rule of three, and may
be stated as follows:—

If x varies as y when z remains constant, and x varies as
z when y remains constant, then x varies as y z when both
y and z suffer change.

Here the quantity of motion varies as the weight, when the
velocity is constant; and varies as the velocity, when the
weight is constant.

∴ the quantity of motion varies as the (weight) × (velocity)

The weight of a body is proportional to the quantity of
matter in it, and we are therefore led to represent the

momentum of a body by the product of the numbers ex-pressing its mass and velocity :

thus momentum = mass × velocity;

or, expressing the mass and velocity by the symbols M and v, we have the relation

$$\text{momentum} = M\,v.$$

This formula cannot be applied to numerical examples until the unit of mass has been selected. It is, however, an advantage to connect a statement of principles with definite experiments, which are easily reproduced, and which con-firm the statement. In illustration of the first clause of the law, the following apparatus may prove useful.

Two light brass wheels, A and B, with axes terminating in conical pivots, are grooved to receive a silk cord, and turn with as little friction as pos-sible.

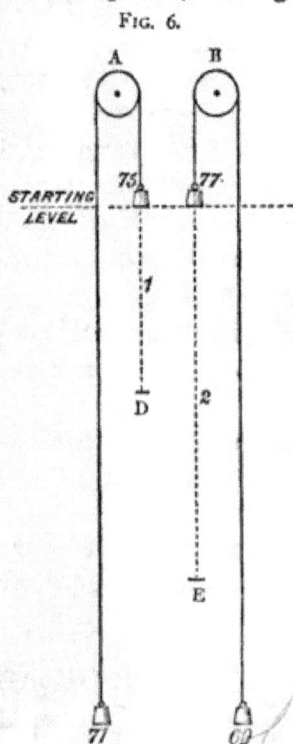
Fig. 6.

To the cord passing over A the weights 71 and 75 units are at-tached, while on that passing over B the weights 69 and 77 units are suspended.

Now $71 + 75 = 146 = 77 + 69.$

Also $75 - 71 = 4$

$77 - 69 = 8$

Hence the mass moved is in both cases 146 units, while the force is the pull of the earth on 4 units in the one case, and that on 8 units in the other case. The forces are as 1 to 2, the masses moved are constant.

If the weights 75 and 77 are started together by supporting them on a small piece of board, and suddenly lowering it out of the way, it will be found that the weights will strike together on two little slabs D and E placed at distances 1 and 2 from the starting level.

This experiment confirms the law that the velocity generated in a given mass is in direct proportion to the force which produces it, for we reason as follows :—

If, in every small interval of the motion, the force 2 generates in the constant mass twice as much velocity as that generated by the force 1 in the same mass, then the whole space which one weight is carried through in any given time must be twice the space through which the other weight is carried in the same time.

By varying the experiment we may show that the velocities generated in any given time are as 1 to 2. Provide two rings, as well as the tables, and support them in the path of the descending weights ; let the extra weights 4 and 8 be elongated strips of metal which will rest on the rings as the weights pass through : this contrivance is due to Atwood ; while the arrangement of the double observation, which has been described, is that of Professor Willis.

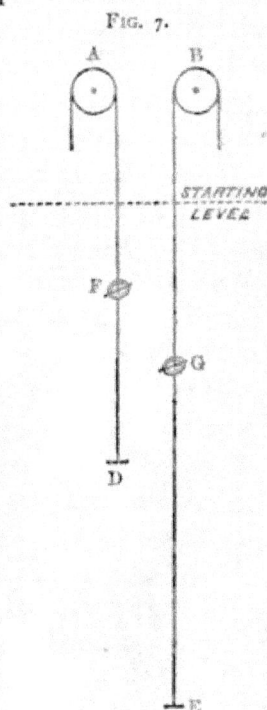

FIG. 7.

Now start the weights as before ; place the rings at F and G, which are at distances 1 and 2 below the starting level, and the tables D and E at further distances 1 and 2 ; the weights will strike D and E at the same instant. The moving weights 4 and 8 will be left on the rings at the same instant, and the masses being then balanced will move on uniformly with the velocities acquired. Since the space F D is one half the space G E, it is evident that the velocity acquired in falling to F is half that acquired in falling to G.

Our conclusion is that when the mass moved is constant, the velocity generated in any time is in direct proportion to the force which acts upon the mass.

As regards the weights, or the units of weight selected, no restriction is necessary, and they may be varied at pleasure, but they must be suited to the scale on which the apparatus is made.

In order to show that when the velocity remains constant the force required to produce that velocity varies directly as the mass moved, we may suspend over the pulley A the weights 71 and 75, and over the pulley B the weights 142 and 150. The masses moved will now be 146 and 292, whereas the moving forces will be 4 and 8. If we place the tables D and E at the same distance below the starting level, the weights will be heard to strike with one blow. The force 8 acting on a mass 292 is equivalent to the force 4 acting on 146.

In this way we learn to conceive the idea of *mass* as a property of matter quite distinct from its weight. It is no doubt true that we estimate the mass of a body by its weight ; but we can fix our minds on a piece of matter travelling through empty space and quite divested of the quality of weight. The weight of this piece of matter may have gone, but its mass will be precisely the same as if it were resting quietly on the earth's surface. So long as matter exists, and we believe matter to be indestructible, the property which we express by saying that matter possesses *inertia*, remains the primary property which is ever one and the same and unalterable. The power of matter to resist the action of force is believed to be inseparable from matter of every kind and wherever situated.

The second clause of the law asserts that the change of motion takes place in the direction of the straight line in which the force is impressed.

Hence if a ball be thrown in any direction across the deck of a ship moving uniformly, there will be nothing to suggest that the ship is in motion so long as we refer the movement only to points in the vessel itself. A ball dropped in a railway carriage in rapid motion will appear to describe a vertical path just as if the carriage were at rest, the change of motion due to the attraction of the earth taking place in the vertical direction in which this force necessarily acts. So again we are unconscious of the rapid movement which sweeps us onward through space : everything appears to be' in repose, because the forces which we impress upon bodies produce their full effect in the lines of their proper action precisely as they would do if the earth were motionless.

A direct experiment is the following :—

Two balls A and B are started at the same instant ; one falls vertically through A M, and the other is projected horizontally in the direction B T. Both balls will strike a horizontal plane through any point M at the same instant. Here the attraction of the earth acts upon A at rest, and upon B in motion ; but it pulls each body through precisely the

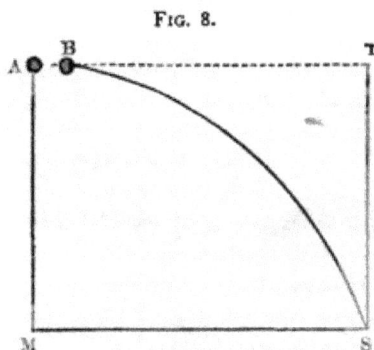

FIG. 8.

same vertical space in the same time. Although A travels in a direct line A M, and B in a curvilinear path B S, the vertical movement T S of the projected ball is equal to A M, the path described by the ball which drops from rest. Our law of motion would have enabled us to affirm this result beforehand, and the conclusion forms the basis of the theory of projectiles.

It may be objected that the illustrations given in confirmation of a law which refers mainly to a change in motion already existing, have been supplied by regarding the motion set up in bodies which we should describe as being at rest. The answer is, that no experiment can be made on a body at rest, for no such thing can be found. If we take up a bullet, it appears to be at rest in the hand, but it is actually sweeping through space with a velocity far greater than that impressed upon it by gunpowder when it is fired from the barrel of a rifle. Any result that we can obtain in a body said to be at rest is really a result obtained in a body already in the most rapid motion, and it must not be forgotten that these illustrations do not prove a law of nature; they merely suggest or confirm the probability that it is true. When the student has toiled up to a mastery of his subject, he will find that the proof of these laws of nature rests upon the continual and accumulated evidence of agreement between observation and theory, that it is based upon the prediction of results which have been foreseen by the aid of the laws, upon discoveries such as that of a planet whose presence has become suspected because its power has been felt.

ABSOLUTE UNIT OF FORCE.

These two laws of motion supply us with a definition and measure of force.

Definition. *Force is any cause which changes or tends to change the motion of a body by altering either the quantity or the direction of that motion.*

It will be observed that there is no mention of a pound weight in defining force, and that no support is given to the erroneous notion that a pound is a force.

The measure of a force when uniform is the quantity of motion which it produces in a unit of time, hence,

Definition. *Force is measured by the velocity generated in one second in the unit of mass in a free body.*

For considerable forces we select the standard pound as the unit of mass, and the unit of force will then be '*a force which acting on one pound for one second generates in it a velocity of one foot per second.*'

This unit is named the *absolute unit of force,* because it

applies everywhere, is quite independent of locality, and is derived from the one unalterable property of mass.

Where the **forces are** minute, as in e'ectrical measurements, **smaller units are** selected, thus the unit of force given in the 'Text-book on Electricity' is a force which will produce a velocity *one centimètre* per second in a free mass of *one gramme* by acting on it for one second.

The method of measurement is still that of absolute measure whether the units be pounds, feet, seconds, or grammes, centimètres, and seconds.

Forces measured in terms of the absolute unit are said **to** be expressed in *absolute measure.*

MEANING OF THE LETTER g IN MECHANICS.

A great number of experiments have been made from time to time in order **to** ascertain the exact velocity generated in a body falling under the attraction of the earth and freed from the resistance of the air. The conclusion arrived at is that the standard pound weight so falling in London would acquire a velocity of 32·1889 feet per second. This number varies with the latitude, according to an ascertained law. In latitude 45°, **near** Bordeaux, for example, it has the value 32·1703.

In our books **on mechanics the letter** g is appropriated for representing the **velocity acquired in one** second by a body *falling from rest* in a space devoid **of air.** This is the primary meaning of the symbol g, but there are other secondary meanings, equally important, which we shall endeavour **to make** clear. For England the numerical value of g is **taken to** be 32·2.

GRAVITATION UNITS OF FORCE AND MASS.

We have spoken of the gravitation measure as that wherein a force **which** will support w pounds of matter is called a force of w pounds ; and we now define the units **of** force and mass in order that the student may be able to compare the absolute and gravitation measures.

Def. *In gravitation measure, forces are measured by the weights they will support.*

The *unit* is a pound weight, whence a force which will support 3 pounds is represented by the number 3.

Def. *In gravitation measure the unit of mass is the quantity of matter in a body weighing g pounds.*

This is an arbitrary assumption, liable to the objection that the unit changes with the locality, but made for very cogent reasons, which will be discussed presently.

Let M be the mass of a body weighing w pounds. Since a body of mass 1 weighs *g* pounds, we infer that a body of mass M weighs M*g* pounds, or that

$$w = M g, \text{ and } M = \frac{w}{g}.$$

It is easy to express any force in gravitation measure when we know the velocity which the force will generate in one second in a body of given weight.

For example, let a force F generate a velocity of 47 feet in one second in a body weighing 5 pounds.

If the pull of the earth, or the weight of 5 pounds, were to act on the body, it would fall and acquire a velocity of 32·2 feet in one second. The force then acting would be called a force of 5 pounds, and we shall compare the forces F and 5 by comparing the quantities of motion produced by each in one second.

Hence
$$\frac{F}{5} = \frac{\text{mass} \times 47}{\text{mass} \times 32\cdot2} = \frac{47}{32\cdot2},$$

$$F = \frac{5 \times 47}{32\cdot2} = 7\cdot3 \text{ pounds.}$$

Generally, if a force F generate a velocity *f* in one second in a body weighing w pounds, we have

$$\frac{F}{w} = \frac{\text{mass} \times f}{\text{mass} \times g} = \frac{f}{g}, \quad \therefore \quad F = \frac{wf}{g} \text{ pounds.}$$

1. *In absolute units, the force of attraction of the earth on the unit of mass,* that is, the standard pound, *is expressed by* 32·2 *or g.*

This appears from the definition. A force which will generate a velocity g feet per second is g times as great as a force which will generate a velocity of 1 foot per second. But the pull of the earth generates the velocity g in the standard pound in one second; hence the pull of the earth on the standard pound is g times the unit of force, or is expressed by the number g.

2. *In gravitation units the pull of the **earth on the** standard pound is expressed by the number unity.*

Hence we pass from the gravitation to the absolute measure of force by multiplying every so-called force of a pound by the number g; and we pass from absolute to gravitation measure by dividing every number representing a force by g.

For example. In absolute measure the *unit of force* is $\frac{1}{32·2}$ of the attraction of the earth on a standard pound, and may be represented very roughly in gravitation measure by about half an ounce weight.

In like manner we compare the expressions for the momentum of a body according to the two measures.

Let a body of weight w be moving with a velocity v.

In *gravitation units* we estimate its momentum by $\frac{w\,v}{g}$, whereas in *absolute units* the same momentum is w v.

The measurement of force has been discussed; and it may be well to point out that if we were at liberty to interpret the laws of mechanics without any reference to previous writers, and to disregard the modes of expression adopted by engineers, we might estimate all the forces with which we have to deal in absolute units and abandon the gravitation measure. But it would be extremely inconvenient to do so, and it would be very embarrassing if we were to ' adopt sometimes one measure and sometimes the other.

The most convenient course will probably be to adhere to the measurement of force by pounds, and to give all our results in the usual manner. The change to absolute units is perfectly easy, and can be made in a moment by simple multiplication. The student will therefore possess a complete command over both systems while putting only one into practice.

THE MOTION OF A FALLING BODY.

Before discussing the third law of motion, it will be necessary to obtain some knowledge of the laws which govern the motion of a falling body.

The first fact observed, is that all bodies, whether light or heavy, fall to the ground at the same rate when freed from the resistance of the air.

That this ought to happen may be shown as follows :—

Let the dots in fig. 9 represent a set of molecules of equal size and weight, not connected with each other. If they be allowed to fall, they will all move downwards at precisely the same rate, and preserve always the same relative positions. Hence, when separate and unconnected, the attraction of the earth will impress upon each molecule the same quantity of motion that it would impress if they were rigidly linked together. That being so, it is manifest that a rigid body made up of molecules falls at the same rate as its separate particles would fall if they were disconnected ; and this is true, whether the particles be condensed and closely packed, as in a piece of metal, or expanded and fewer in number, as in a feather.

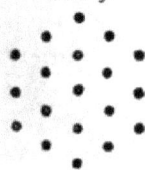

FIG. 9.

Accordingly, there is the old experiment of the guinea

and a feather which fall to the bottom of a tall exhausted receiver at the **same** rate.   The fact **may be** tested without
an air-pump, by placing **the feather on a** large coin, such as
a half-crown.   The coin must **be** carefully dropped **with** its
face horizontal, so that it does not turn **over,** the feather **will**
then be unaffected by the air, **and** will **reach** the ground at
the same instant as the coin.   So two equal balls, one of
lead, the other of wood, will fall to the ground in a time
which is sensibly the same.   The experiment would succeed
equally well if the wooden ball were replaced by a light
hollow ball of india-rubber.   This result may appear to con-
tradict the observation, that when we employ artificial force
to move a body, **the** velocity generated depends **upon the**
weight **of the body moved.**   Thus, if a given **force** generates a velocity of **one** foot per second in a body of one pound
weight, it will generate $\frac{1}{10}$th part of that velocity in a body
of ten pounds weight.   But the force of attraction of the
earth, which is a very different thing from the pull of a
string, will generate precisely **the** same velocity in both
cases.

The explanation is, that **the** quantity of attractive force of
the earth increases **in direct** proportion to the quantity of
matter acted on ; whereas, **any** artificial force, such as the
tension of a string, depends on the source from which it is
derived, **and** does not change with the body on which it is
acting.   The observed fact, when understood, is an additional
confirmation of the truth **of** the laws of motion.

We pass on to discuss the formulæ which determine the
motion of a falling body.

Suppose a body to *fall from rest* in a space freed from air
or other resisting medium.   It is found by experiment that
it will acquire in one second a velocity of **32·2** or **g** feet per
second.

Let v be **the velocity of the** falling body at the end of t''.
Then the attraction **of the** earth would impress an additional
velocity g in the **second** interval of $1''$, and would do the

same in each successive interval, whereby the velocity acquired in t seconds would be gt. Hence

$$v = gt, \quad . \quad . \quad (1)$$

This is the first fundamental formula whereby we connect the *velocity and the time of falling from rest.*

The student will now be asked to consider a geometrical method of computation which was given to us by Newton, and has many applications.

Before discussing it we may show that the formula $v = gt$ can be expressed geometrically. For this purpose take the line A K to represent t seconds, and let K L, at right angles to A K, represent the velocity acquired by a body when falling from rest during t seconds.

FIG. 10

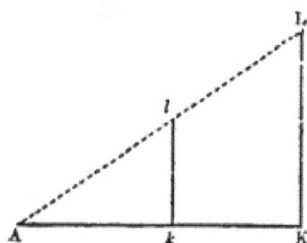

Then $\dfrac{K L}{A K} = \dfrac{v}{t} = g.$

If A k, $k l$ be any two other corresponding values of the time and velocity, we have, as before,

$$\frac{k l}{A k} = \frac{\text{vel.}}{\text{time}} = g = \frac{K L}{A K}.$$

But this proportion can only be satisfied when the point l lies in A L, therefore A l L is a straight line. We now construct for the velocity acquired in any given interval of time (such as A k) by joining A L and drawing $k l$ parallel to K L.

Prop. To find the number of feet through which a body falls from rest in t seconds.

Conceive that the time is divided into a number of extremely small intervals, all equal to each other. The artifice employed by Newton consisted in substituting the aggregate of a series of step by step movements, each uniform in itself, for the continuous motion of the falling body. It is evident that a minute additional velocity is imparted during each minute interval of time, and we shall proceed to estimate the space described on *the hypothesis of a movement by steps.*

Conceive that the time A K is divided into a number of equal portions, two of which are represented by *h k* and *m r.*

Since *h k* represents a very minute interval of time, we may suppose the body to move uniformly during the time

FIG. 11. FIG. 12.

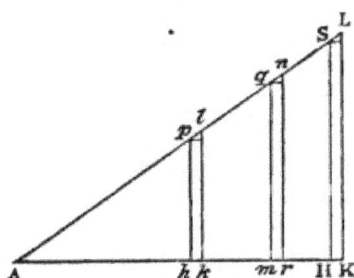

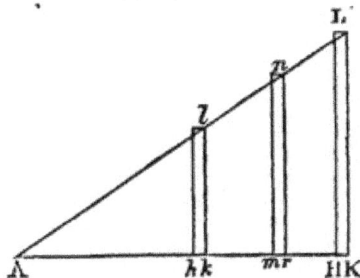

h k, (1) with the velocity which it has at the beginning of that interval, (2) with the velocity which it has at the end of the interval. Complete the parallelograms *p k* and *h l* in the two diagrams, then the space described in time (A *k*—A *h*) will be represented either by *p k* or *h l* according to the hypothesis made.

Similarly the space described in time (A *r*—A *m*) would be represented either by *q r* or *m n.* Proceeding in this manner, the spaces described would be the aggregate of a series of rectangles whose sum is less than the triangle A K L on the first supposition, and greater than it when the second hypothesis holds good.

It is evident that the space actually described lies between the sums of the spaces resulting from the step by step movements, which we may call (A) and (B). Now the difference between (A) and (B) cannot exceed the rectangle contained by H K and K L, which rectangle may be made less than any assignable quantity by diminishing H K.

Hence we argue as follows.

The *true space* lies between (A) and (B), and so does the triangle A K L, but the difference between (A) and (B) can be made less than any assignable magnitude, therefore the difference between the *true space* and the triangle A K L is less than any assignable magnitude, and therefore the *true space* is represented by the triangle A K L.

Hence the space described in t seconds

$$= \text{area A K L,}$$
$$= \tfrac{1}{2} \text{A K} \times \text{K L,}$$
$$= \tfrac{1}{2} t v.$$

But $v = g\,t$, therefore space described

$$= \tfrac{1}{2} t \times g t,$$
$$= \tfrac{1}{2} g t^2.$$

Let s be the space through which the body *falls from rest* in t seconds, then we have the formulæ

$$s = \tfrac{1}{2} t v, \qquad . \qquad . \qquad . \quad (2)$$
$$s = \tfrac{1}{2} g t^2, \qquad . \qquad . \qquad . \quad (3)$$

which connect the *space from rest, time and velocity*, or the *space from rest and the time.*

Combining the formulæ $s = \tfrac{1}{2} g t^2$ and $v = g t$, we have

$$s = \tfrac{1}{2} g \times \frac{v^2}{g^2},$$
$$= \frac{v^2}{2g},$$

or $\quad v^2 = 2 g s, \quad . \qquad . \qquad . \quad (4)$

which formula gives the velocity acquired in falling through a given space, or connects the *space from rest and the velocity.*

THE UNIT OF MASS IN GRAVITATION MEASURE.

We can now understand why the quantity of matter in a body weighing g pounds is selected as the unit of mass.

The mass of a body is to be a numerical representation of the quantity of matter in it. Our estimate of the mass of a body comes from its weight, and we have therefore to consider whether it is possible so to vary the unit of mass that it shall always increase or decrease in the exact proportion in which the weight of a body varies in consequence of a change of locality.

By choosing g pounds as the unit of mass, this object is effected.

For, let w be the weight of a certain quantity of matter in a certain locality, say in London ; and g the velocity acquired in one second by a body there falling.

Again, let w′ be the weight of the same quantity of matter at the equator, say, of the planet Jupiter, and let g' be the velocity acquired in one second by a body falling at the equator of the planet.

Then the quantities of motion generated in each case are as the weights of the substance on the earth and Jupiter.

$$\therefore \frac{w}{w'} = \frac{mass \times g}{mass \times g'} = \frac{g}{g'}$$

$$\therefore \frac{w}{g} = \frac{w'}{g'}.$$

Hence although both w and g vary with the locality, the ratio $\frac{w}{g}$ does not change, but is the same wherever the same mass of matter is to be found.

It will be very convenient, therefore, to represent the mass of a body weighing w pounds by the fraction $\frac{w}{g}$.

In order to do so it will only be necessary to assume that the unit of mass is the quantity of matter in a body weighing g pounds, and changes in the same proportion as g itself changes. Of course we introduce the defect of a unit of variable magnitude, one unit for London, another for Paris, and so on. The advantage to counterbalance that defect is, that our numerical representation of the same quantity of matter is the same everywhere ; that is to say, a mass represented by the number 10 in London would be represented by the same number 10 on the surface of Jupiter or of the sun.

D

To make this clear, the attraction of the mass of the sun on a body at its surface is about twenty-eight times that of the earth, or a mass weighing 10 pounds on the surface of the earth would weigh 280 pounds if transported to the sun.

According to our statement, a body of mass 1 weighs g pounds ; therefore a body of mass $\dfrac{1}{g}$ weighs 1 pound, and a body of mass $\dfrac{w}{g}$ weighs w pounds.

Let the number 7 represent the mass of a body weighing w pounds here on the surface of the earth, then $\dfrac{w}{g} = 7$.

Conceive that the body is transported to the surface of the sun ; it will be pulled down by the enormously increased mass of the sun, and will weigh 28 w, in the place of w, which before represented its weight. For a similar reason g will become 28 g. Hence the mass of the body when removed to the sun is $\dfrac{28\,w}{28\,g}$, or $\dfrac{w}{g}$, or 7, as at first.

Thus the mass is represented by a fixed number in both cases, as it ought to be represented, for it is clear that the mass is not affected by the removal.

In this way, and for the above reasons, the unit of mass is selected in the gravitation measure of force.

WORK STORED UP IN A RAISED WEIGHT OR IN A WEIGHT
IN MOTION.

Work is done when a weight is raised in opposition to the pull of the earth. This is the simplest idea that we can form of what is meant by work, and is that from which we derive the measure of work.

Generally, we say that work is done in moving a body against a resistance. The work is *done* or *performed* and the resistance is *overcome* by the action of force upon the body moved.

For the present we suppose the point of application of the

force to be moved in a direction exactly opposite to that in which the resistance is acting. This would occur when a weight hanging on a **rope** is lifted by a force acting through the rope.

The *work done* is then measured by the product of the number **of** pounds lifted into the **number** of feet through which they are lifted. **The** product **is said to** be a number of *foot-pounds*.

Ex. If ten pounds be lifted through five feet, the work done = 10 × 5 = 50 foot-pounds.

There is another mode of estimating work derived from the laws of motion which we have now to compare with that already defined.

When force acts upon a free body it will set it in motion, the inertia of the body presenting a resistance which is over-come by the force, and thus work is done in impressing velocity upon a body. So also a body in motion can only be brought to a state of rest by the action of force, and work is done during the destruction of the motion.

If we were to raise a body through a certain height, and then allow it to fall, it would acquire, in falling, a velocity dependent on the height **through** which it had been raised. Conceive that the direction **of** its motion is now suddenly reversed, the weight will rise to the exact height from which it fell, and in doing **so** will perform work. The conclusion is, that **the** velocity acquired in falling is a measure of the work **done in** lifting the body, and is produced by the action **of a definite** force, namely, the weight of the body, acting through **a** definite space, namely, the height through which the body **has** been raised.

But any other body moving with any given velocity might have acquired it in like manner by being subjected to the action of the force of gravity while falling through a deter-minate height, and thus we say that work is stored up in a body in motion, and that the measure of the work is the height through which the body must be lifted in order that by falling it may acquire the velocity with which it is actually moving.

Prop. To estimate the work stored up in a body of weight w when moving with a velocity of v feet per second.

Let h be the height through which a body must fall from rest in order to acquire a velocity v. Then

$$v^2 = 2gh, \text{ or } h = \frac{v^2}{2g}.$$

But the work done in lifting a body of weight w through a height h is wh, hence

$$\text{the work done} = wh = \frac{wv^2}{2g}.$$

For example, let a bullet leave the barrel of a gun with a velocity of 1000 feet per second, and suppose it to weigh one ounce, we should determine the work stored up in the bullet from the formula given above. Here $v = 1000$,

$$\text{therefore } h = \frac{(1000)^2}{2g} = \frac{1000000}{64 \cdot 4} = 15527 \cdot 95 \text{ feet.}$$

Hence we say that the powder has expended as much work as would lift the bullet through the space of 15528 feet through which it must fall in order to acquire that velocity. But the estimate of work is always in foot pounds, and therefore we convert our result into these units.

That is, the work done $= \frac{1}{16} \times 15527 \cdot 95$ foot pounds.

$$= 970 \cdot 5 \text{ foot pounds.}$$

The velocity hitherto considered has been linear velocity, but there can be no motion of any kind, whether it be of translation or rotation, without the action of force in overcoming resistance. The inertia due to mass is ever present, and work can be stored up in a rotating wheel as certainly as it can be accumulated in the ponderous head of a steam hammer. Thus a heavy rotating body, such as the fly wheel of an engine, is symbolised as a reservoir into which the work of the engine can be poured just as water is poured into a vessel.

The work stored up in a body rotating with a given velocity may be thus estimated.

Conceive a body of weight w to move in a circle of radius r with a linear velocity v.

A line drawn from the body to the centre of the circle will rotate round the centre with an angular velocity, which we call ω.

Then $v = \omega\, r$.

But the work stored up in the body is

$$\frac{\text{w}\, v^2}{2\,g} = \frac{\text{w}\, \omega^2\, r^2}{2\,g} = \frac{\omega^2}{2}\ (\text{mass})\ (\text{radius})^2.$$

From this expression we conclude that the work stored up in the separate parts of a body moving with a given angular velocity depends upon the *square of the distance* of each part from the axis. A pound weight at a distance of 3 feet from the axis of rotation has 9 times as much work stored up in it as the same weight at a distance of 1 foot from the axis, the angular velocity being the same in both cases.

MEANING OF THE TERM ENERGY.

The term *energy* is restricted to one particular meaning in mechanics, and signifies the *capacity for performing work*.

A body possesses energy when it is capable of doing work; thus, a raised weight possesses energy, for we know that we can obtain from a falling weight the exact amount of work which was expended in raising it. In this way clocks and small machines are kept in motion for days by the gradual expenditure of work done in raising a weight.

The energy which exists in a raised weight is named *potential energy*. It may or may not be called into action, it may lie dormant for years ; the power exists, but the action will only begin when the weight is released and allowed to commence falling. Hence the word 'potential' is a very significant term, as expressing that the energy is in existence, and that a new power has been conferred upon the weight by the act of raising it.

The use of the term *potential energy* is not limited to the case of a weight raised. It applies to every portion of matter which is at rest, and is nevertheless capable of doing work. It is distinguished by being applied to matter in a state of repose, and contrasts with the phrases *actual energy* or *energy of motion,** which denote the energy and power of doing work that appertains to a body when in motion.

A body in motion, as we have seen, has work stored up in it, which work it must yield up before it can be reduced to rest, and thus we apply the phrase *energy of motion* as expressive of the power existing in every moving body.

There is, however, no limitation to the use of the word energy, in whichever sense we regard it ; and the laws which govern the transfer of energy from one body to another, apply to the minutest portions of matter and to the smallest forces in nature just as certainly as to the movement of a railway train by the pull of the engine.

If there be potential energy in the steam confined in the boiler of a locomotive, so also there is in the coal which is being burnt, for that is the primary cause of the motion ; and now arises the question, what do we mean by the potential energy of a piece of coal.

Conceive that a mass of coal is made up of a number of molecules, mainly of carbon and hydrogen, which have been separated by the action of radiation from the sun, and were combined with other molecules of oxygen in past ages. If the coal were set on fire in the air, the molecules of carbon and hydrogen would again rush towards the molecules of oxygen under the action of certain chemical forces, and the same mechanical result would be obtained as if a weight fell to the ground.

The separated atoms of carbon possess potential energy which is converted into energy of motion during the burning, and is finally rendered up in the form of work. By falling together the molecules acquire motion just as weights

* *Actual energy,* or *energy of motion,* is now commonly called *kinetic energy.*

acquire motion in falling downwards. It is true that the movements of the molecules are of almost infinite minuteness, and cannot be detected as a matter of observation, any more than they can be seen when a straight riband of steel spring is bent into a circle. This difficulty always presses upon us. Molecular motion cannot be seen by the eye, and can only be felt as heat. But we reason upon observed facts, and it is believed that we do not err in applying the statements of Newton to every conceivable instance of matter in motion, whether the matter be a molecule or a pound weight.

No substance can burn without evolving heat, that is, without having additional motion impressed upon its separate parts, or in other words without the conversion of potential into actual energy; and the additional motion or heat so developed, is energy available for transformation into mechanical work.

The mere fact that a mass of matter is inert, and apparently quiescent, does not reveal to us its real condition. It may yet be a source of enormous power. A mixture of pounded sugar and chlorate of potash is a simple white powder, apparently powerless, but touch it with a drop of sulphuric acid and an intense exhibition of light and heat is the result. Thus a chemist regards many substances as powerful in a sense quite distinct from that in which the word is accepted in mechanics. In dealing only with masses, we should apply the term 'potential energy' to that source of power which depends upon the position of a body at rest and not to that which is inherent to the position of its separate molecules. It is, however, of importance that the distinction should be understood.

The student is now in a position to understand the third law of motion as given to us by Newton; and it will be found that this statement, when we attach to its terms that wide and extended meaning which modern science demands, does in truth embody the great principle of the conservation of energy.

Third Law. *To every action there is always an equal and contrary reaction, or the mutual action of any two bodies are always equal and oppositely directed.*

This law is sometimes stated as follows. *Action and reaction are equal and opposite.*

The word 'action' may signify *pressure*. If we press the hand upon a table, the pressure will be resisted, the hand exerts an action upon the table, and the reaction which is equal and opposite to the pressure exerted, may be felt where the hand rests. Here action and reaction are both pressures, and no motion results.

Again, the word 'action' may signify *quantity of motion.* In that case, the quantity of motion constituting the reaction will be exactly equal and opposite to the action. When a cannon-ball is fired from a gun, the recoil of the gun exhibits a quantity of motion exactly equal to that of the ball. That this recoil is a force capable of doing work is well known; and it will be seen hereafter that it is nothing else than a form of energy which may be made subservient to useful purposes.

But if a quantity of motion can be regarded as a form of energy, there is surely no reason for restricting these terms 'action' and 'reaction' within mere narrow limits, when, by extending them we can grasp some of the more subtle phenomena of nature.

Finally, therefore, the word 'action' may signify *energy*, and reaction may mean the same thing, and thus, whether we deal with energy as *potential*, or as *actual*, that is *kinetic*, this third law expresses in a few brief words the great principle of the indestructibility of energy. It has long been believed that matter is indestructible, but this belief has not until recently been extended to energy. In the time of Newton it was supposed that the motion arrested by friction was absolutely lost and put out of existence. In later days it is quite common to find in the text-books on mechanics a so-called proof that in the impact of imperfectly elastic bodies work is lost, or that there is less energy in existence after the impact than there was before it. Such statements are of course entirely baseless.

There is no destruction of motion by friction; the movement of the mass is replaced by that of the individual molecules, and we do but pass by a rapid change from that motion, which is seen by the eye of sense, to a new movement which is equally apparent to our higher faculties.

There is not any destruction of energy by impact or otherwise; and whenever motion ceases, whether it be sensible and that of a mass, or insensible and that of its molecules, new positions are taken up: potential energy supplies the place of energy of motion, and the interchange is in no sense a destruction, it is not even a diminution of the source in the smallest conceivable degree.

This is the principle of the conservation of energy. As potential energy disappears, kinetic energy comes into play, and the sum of these energies throughout the universe is constant. To create or destroy energy is as impossible as it is to create or annihilate matter.

ILLUSTRATIONS AND EXAMPLES.

According to the method of instruction proposed for adoption in this book, the statement of principles will be followed by a notice of certain useful applications, and a few easy examples will be worked out or suggested.

The study of mechanics requires the most extensive observation, for it often happens that a principle is applied in a new manner to some useful purpose, notwithstanding that men have been aware of the truth of the principle for centuries. Mechanics cannot be learnt from books alone. The student must go out into the world, and see how mechanicians or engineers accomplish what they have to do, he must continually reason upon what he sees, and, retaining a firm hold of mechanical principles, he may thus gradually obtain a knowledge and mastery of his subject.

Taking the first law of motion, which asserts the inertia of matter, we may notice many useful applications.

THE CARRYING OF CORN ON BANDS.

In the corn warehouses at Liverpool, the grain is carried upon a plain flat band 18 inches broad and made of canvas and india-rubber. This method of transport is found to be extremely economical, the power absorbed being about $\frac{1}{18}$th of that expended according to the old process, where a screw pushed on the grain by rotating in a hollow casing.

The speed is limited by the action of the air, which blows off the grain when a certain speed is exceeded; thus oats may be carried at the rate of 8 feet per second. It is really the inertia of the air which sweeps off the grain, though it is commonly said that the grain is blown off.

The band runs upon rollers, and the point to which attention is now directed, is the method of diverting the grain from one path into another during its passage. The first law is here applied very ingeniously. At the point where the change of path occurs, the carrying-band is bent a little upwards, as shown at B in the sketch.

FIG. 13

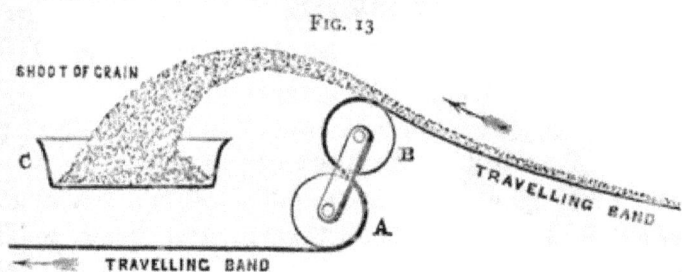

This bending is effected by means of a moveable carriage supporting the wheels A and B, which can be placed anywhere, and the result is that the stream of grain retains the velocity which is given to it by the band, and is carried forward in a jet over the top of B, just as if it were a stream of water. On looking down upon the corn at B, it is quite remarkable to notice how the separate grains retain their relative positions, and shoot forward as if they were glued together. The spout C diverts the corn into a new channel, and may pass it on to another travelling-band for transport in a new direction.

If it be required to deposit the grain upon a travelling band, it will be necessary to give its particles a horizontal velocity equal to that of the band. For this purpose, the grain is directed down an inclined shoot, and acquires thereby a horizontal velocity equal to that of the band ; it is therefore at rest relatively to the band as soon as it comes upon it, and does not flow over the edges.

As a last precaution, the edges of the band are bent up by oblique rollers so as to form a hollow trough where the grain is deposited, and the vertical velocity which the grain has acquired in falling down the shoot is thus prevented from causing an overflow in a lateral direction.

THE INERTIA OF A RIFLE-BULLET.

Another example of inertia is seen when a cylindrical leaden bullet is fired from a grooved rifle. In this case the bullet is expanded or *upset*, as it is termed, by the explosive force of the powder, and the lead is driven into the grooves of the barrel so that the bullet becomes moulded to the bore. It is actually more easy for the powder to distort the bullet, and compress it into the grooves, than it is to move it. The bullet fits the barrel quite easily, and would yield to a touch, but a sudden force finds it inert, and can compress it out of shape, though there is nothing for the bullet to rest against.

In confirmation of this remark, it is well known that slow-burning powder will not upset a bullet. The action depends entirely on the suddenness of the blow.

In the old Enfield bullet a wooden plug was inserted in the base in order to assist this expansive action.

It may be asked, how can the bullet be caught without

injury after it is fired? Of course it must be directed against some soft and yielding substance, and when fired into bran, a rifle bullet will penetrate 8 or 9 feet, but will retain its original form without a mark. The bullet will be deformed when fired into flour or saw-dust, substances which, before trial, would appear to be as suitable for the purpose as bran. This is a good instance of harmless absorption of the work stored in a mass when moving with a destructive velocity.

Again, suppose that a hollow shell of iron is fired from a 9-pounder gun with a given charge of powder, and then that a like shell filled with lead is fired from the same gun, at the same elevation, and with the same charge of powder. The loaded projectile will range much farther than the empty one. Why is this? Manifestly because the inertia of the heavier projectile has detained it longer in the gun, the powder has had a longer time to act, is more perfectly consumed, and the work done upon the shell is increased.

THE DISINTEGRATING FLOUR MILL.

Hitherto, in all pulverising apparatus, such as ordinary millstones, edge runners, stampers, &c., the object to be broken up is pressed between two surfaces, and it is not very obvious that grinding can be done in any other way than by employing some contrivance analogous to the pestle and mortar. But *inertia* is a primary property of matter, so that if you throw a grain of corn into the air and strike it a very sharp blow, the grain may be broken up into fragments although it is not supported against anything. Suppose that an iron rod were to strike the corn with a velocity of, say, 100 miles an hour, the rod will encounter the substance in mid-air, and will find it as inert and unyielding as if it were resting on a stone; a few blows struck at a high velocity in rapid succession will break up the separate grains into a mass of powder, and thus wheat may in effect be ground without using any millstones or rollers.

The *disintegrating flour mill*, as it is termed, consists of two circular discs, *rotating in opposite directions* on the same line of shafting. In the drawing A A represents one disc, and B B represents the other disc as seen in section; the discs are a few inches apart, and are furnished with rows of short projecting bars a, b, c, d, e, and h, k, l, m, arranged in concentric rings and studding the surfaces. In one machine the outer ring of beaters is 6 ft. 10 in. diameter, the linear velocity of a beater in the ring being 140 ft. per second, or, about 100 miles an hour. This velocity corresponds to 400 revolutions of the disc in one minute.

The corn enters by the pipe E and is delivered into the space round the shaft D, which carries the disc B. Thence it is

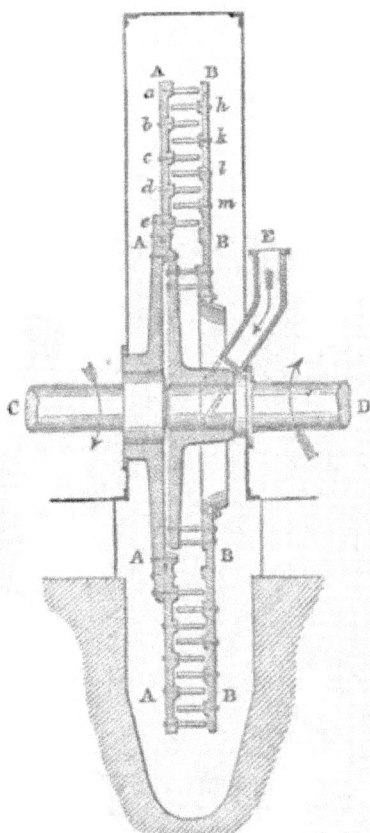

FIG. 14.

carried by the rush of air into the space between the beaters and is struck innumerable blows, until it finally escapes round the periphery in the form of flour. The two shafts C and D revolve in opposite directions, and it will be noted that a grain rebounding from one beater will instantly encounter another whirling round in the opposite direction, whereby the effect due to inertia is heightened. As the material approaches the condition of flour, the particles are reduced in size, and the blows must be more decisive in order to complete the process. This heightened intensity in the blow is supplied by the higher linear velocity of the successive rings of beaters in the passage outwards, for, as we have stated, the energy existing in a moving body depends, not on the velocity simply, but on the square of that velocity.

We shall lose no opportunity hereafter of pointing out that *air and water possess inertia, just as much as solid bodies*, and the truth of this remark would be forced upon any one who examined the disintegrator. When the machine was grinding corn at 400 revolutions per minute the power expended was found to be that of 123 horses ; upon driving the machine without putting any corn into it, the power expended was that of 63 horses. One disc was then disconnected from its shafting, and lashed securely to the other disc, the two discs rotating in the same direction ; the power then consumed was that of 19 horses. This latter power was required to overcome the friction of the machine and the resistance of the external air.

It appeared therefore that a power of (63—19) horses, or of 44 horses, was consumed in churning the air between the discs, i.e., in overcoming the resistance set up in the first instance by the inertia of the air. This result is the more remarkable as the mass of air enclosed between the discs weighed only 2 lbs. If the power of 44 horses can be consumed in churning 2 lbs. of air, it is very clear that air must possess inertia.

But we have also pointed out that heat is motion, and it is of course impossible to knock about a mass of air in this way without heating it. Accordingly, in one experiment the machine was driven empty at 700 revolutions per minute, and in about three minutes the temperature of the casing rose from about 60° to 110° Fahr., although there was a free passage of air between the discs.

Machines constructed on this principle have been in use for many years in pulverising or granulating mineral sub-

stances, especially those which are liable to get into a pasty
condition when crushed.

No doubt the loss of power caused by the action of the
enclosed air is a serious objection in practice, and the
general problem of grinding corn presents so many difficulties that the mechanical features of the machinery form
the only subject-matter for our consideration.

PENETRATION BY A SHOT UNDER WATER.

Sir Joseph Whitworth has recorded some experiments
which illustrate the inertia of water.

In 1868 a hexagonally rifled 3-pounder steel gun was laid
at an angle of 7° below the horizontal line, and a flat-headed
projectile was fired at a target dipping into water. The
point aimed at was 2 feet below the surface of the water,
and the line of aim passed through 6 feet 8 inches of water.
The shot went straight on in the direct line of fire, notwithstanding the water, and as there was no deviation,
there was little to suggest the action of inertia.

But when a projectile somewhat pointed or egg-shaped
at the head was fired from the same gun, it would scarcely
enter the water at all, and struck the target 2 feet above
the water-line instead of 2 feet below it, rebounding much
as if it had been fired against a table of rock.

Here the inertia of water was palpable enough, but the
reason why the form of the head makes so great a difference
in the result is rather beyond our scope at present.

What we have to say now is that the same property of
mass exists in the water in both cases, and that nothing else
than the difficulty of suddenly moving matter can cause a
hard and heavy piece of iron to glance off from a yielding
substance like water.

THE SUPPLY OF WATER FOR TRAINS WHILE RUNNING.

The *inertia of water* was taken advantage of by Mr. Ramsbottom in his well-known arrangement for supplying locomotive tenders with water while the train is running.

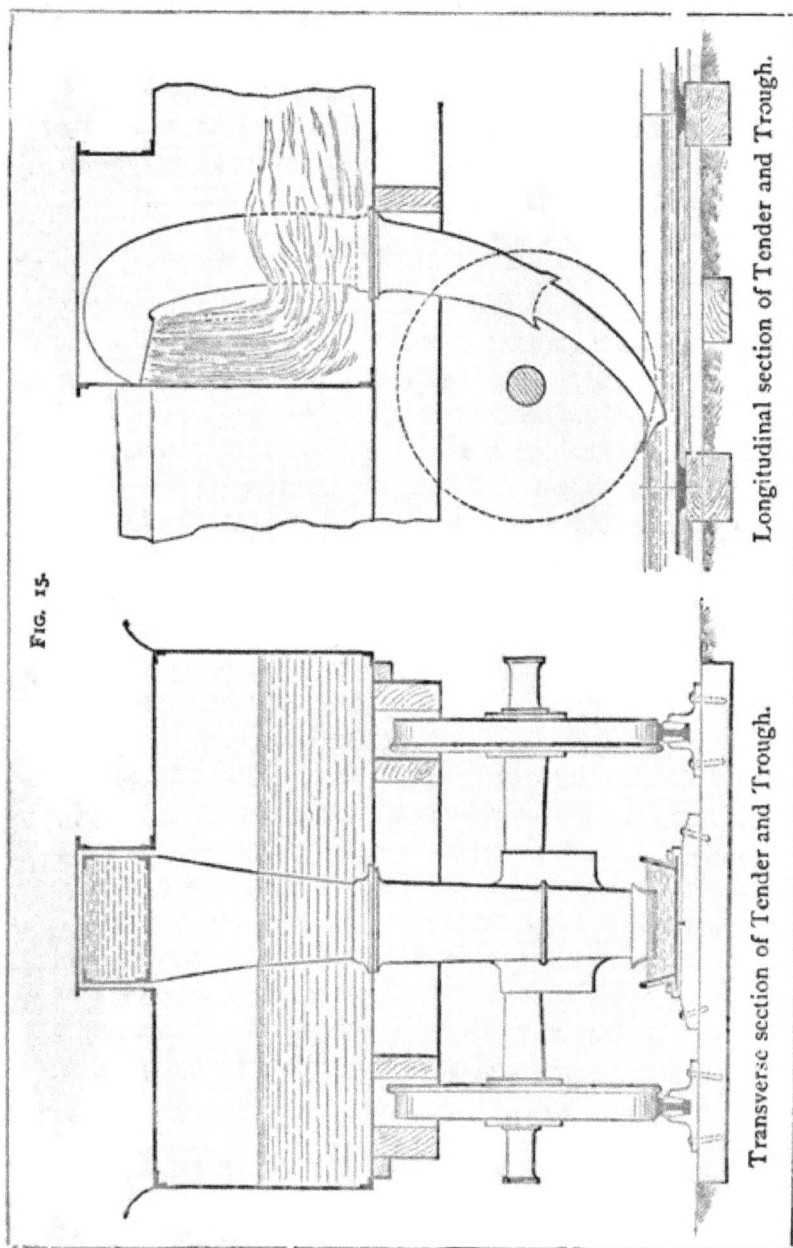

FIG. 15.

Longitudinal section of Tender and Trough.

Transverse section of Tender and Trough.

The Irish mail runs from Chester to Holyhead, a distance of 84¾ miles, in two hours, and the tender picks up about 1,000 gallons of water from a long trough, 18 inches wide and 6 inches deep, which is laid for a length of 441 yards near to Conway. A scoop, 10 inches wide, dips 2 inches into the water, and is connected with a pipe leading to the tender. As the engine runs along, the mouth of the scoop slices off a mass of inert water, and the liquid, before it has had time to acquire the velocity of the train, slides up the few feet of pipe leading to the tender, and rushes into the tank as if it were being discharged from a most powerful force-pump. What really happens is the exact contrary of what appears to happen : the water is at rest, except so far as the movement in a vertical direction is concerned, but an inclined plane is pushed underneath it with a velocity of some 40 miles an hour, and the water is lifted into the tender.

If the explanation be correct, the contrivance would cease to act when the velocity of the train was sufficiently reduced. The height of the discharge orifice is 7½ feet from the ground, and at 15 miles an hour no water gets into the tender, whereas at 22 miles an hour the delivery is 1,060 gallons per run ; this shows the influence of velocity. At 50 miles an hour, the delivery is practically the same, for you can but slice off and seize hold of a layer of water of a given depth at whatever rate the train may be travelling.

FURTHER EXAMPLES OF THE INERTIA OF WATER.

Again, the inertia of water becomes very apparent in pumping apparatus. A force pump was employed in one case to force water through a valve which only rose 1¼ in. at each stroke of the pump. As the valve descended on to its seat after the stroke, the column of water came with it, and the practical result was that the pipe burst more than once near the valve ; on examination, it was found that the blow struck by the column of water returning through

E

this trifling distance, raised the pressure from 36 lbs. on a square inch of the pipe to 156 lbs. at the moment after the valve had closed.

In obtaining a water supply for Manchester, the same property is applied usefully. Here the water is drained from the moor-land lying between Manchester and Sheffield, and is brilliant and pure in dry weather, but becomes discoloured by the peat after rain. The problem is to prevent the tur-bid from mixing with the pure water. The means adopted are quite simple, each stream separates itself of necessity, so that the pure water flows into one set of reservoirs, and the turbid water into others where it can become clear.

FIG. 16.

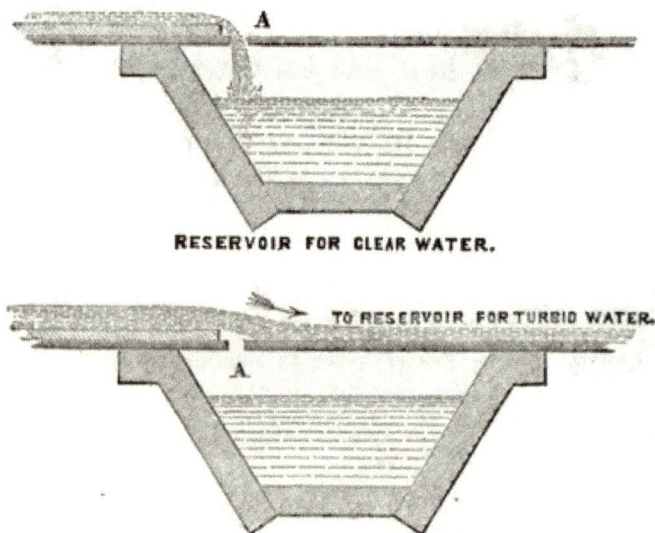

RESERVOIR FOR CLEAR WATER.

TO RESERVOIR FOR TURBID WATER.

The sketch shows the arrangement adapted for a small stream which flows over a ledge having an opening at A. When the water is sluggish and pure, it drains through the opening and falls into the clear-water reservoir; when the stream is swollen by rain, the inertia of the water causes it to leap across the gap, and to pass in another direction.

Something very like this occurs in the making of shot.

It is well known that shot are formed by allowing melted lead to fall in drops from the top of a high tower. The liquid drops become round and solid as they fall and are received into a cistern of water.

The perfectly spherical shot are then separated from the imperfect by allowing them all to roll down a very smooth inclined slab of iron. Those that are perfectly round acquire a velocity sufficient to carry them over certain pitfalls in the way, whereas any defect in shape will cause the imperfect shot to move more slowly, and to drop into the pitfalls, just as the slowly flowing clear water drops into the reservoir designed for it, while the rapid turbid water leaps over the opening.

THE TENDENCY TO MOVE IN A STRAIGHT LINE.

The inertia of matter is felt when a stone is whirled round in a sling, the strain upon the string is due to this property.

Conceive that a smooth ring P is threaded upon a smooth rod A B, centred at A, and capable of whirling round in a horizontal plane.

FIG. 17.

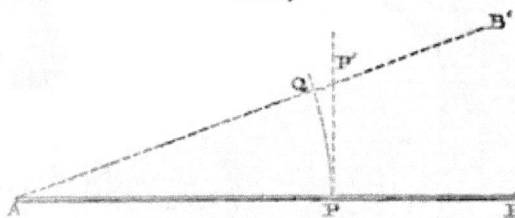

Draw P P' perpendicular to A P; then if A B be moved into the position A B', the ring will be started in the direction P P', and tends to move in that line. At each successive instant the ring receives a pressure in a direction perpendicular to the rod, and continually tends to get farther off from A: thus the ring appears to travel down the rod, and soon slips off at the end.

If P be kept in the circle P Q, it must be pushed towards A, that is, a force must be exerted in order to make P describe a circle round A.

E 2

The motion in a circle under the action of a constant force will be examined hereafter ; at present it suffices to say that the ring would be whirled off the stick by reason of the tendency of matter, when in motion, to continue its course in a straight line with a uniform velocity.

THE VENTILATION OF COAL MINES.

The method of ventilating coal mines first adopted has been by means of furnaces kept burning at the bottom of a shaft. In the *Seaham Colliery*, for example, there are two shafts, one for the air to descend and to pass through the workings of the mine, the other for the ascent of the air at the close of the circuit. The depth of each shaft is about 510 yards. Fires are kept burning at the bottom of the upcast shaft, which is, in fact, a chimney ; and more than 20 tons of coals are consumed every twelve hours in producing the necessary draught. This method is quite effective, but it is certainly not economical. In the place of these furnace fires, ventilating fans are now being introduced, the most

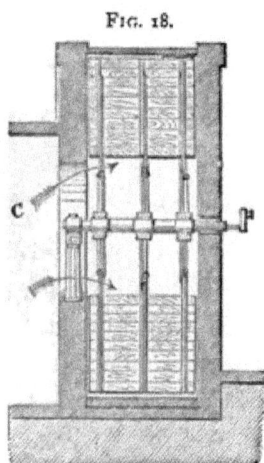

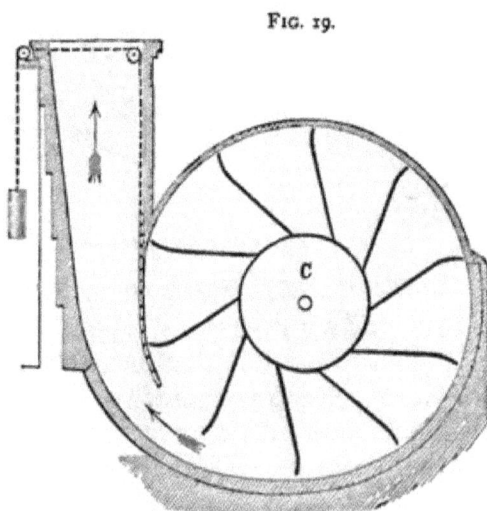

FIG. 18.

FIG. 19.

successful of which appears to be that invented by M. Guibal of Belgium. This mechanical apparatus may be made of a large size, say, 36 feet in diameter, and 12 feet in breadth, making perhaps 60 revolutions per minute. It consists of a large wheel, with a central opening for the entrance of the air, and a number of radial shutters, each corresponding to A B, in the last article. As the wheel rotates the masses of air between each pair of shutters slide out from the centre just as the ring P slides along A B, and the mine may be ventilated as effectually as it could be by a furnace fire. It is a very striking thing to stand in the gallery or passage which feeds the centre of this gigantic revolving wheel with air coming from the mine. There is a strong gale of wind apparently blowing past, and at first it is difficult to believe that the revolving wheel is the sole agent which produces it.

Again, at Liverpool there is a railway tunnel 2,025 yards long and 430 square feet in sectional area. This tunnel is now ventilated by a fan 29 feet in diameter and $7\frac{1}{2}$ feet wide. The fan has twelve straight vanes or arms, made of Bessemer steel, and set in lines radiating from the centre. Each vane is $7\frac{1}{2}$ feet wide and 7 feet long. Here then is the apparatus corresponding to the rod and the ring. The wheel makes 45 revolutions per minute, and is capable of clearing out every particle of the trail of steam and smoke left by a passing train in about 8 minutes. In doing so it is estimated to drag through itself about 115 tons of air.

FIG. 20.

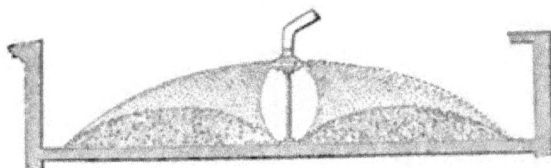

In a similar manner the corn at the Liverpool warehouses can be spread nearly uniformly over a circular space 45 feet in diameter, by means of a fan placed $9\frac{1}{2}$ feet above the floor, and making 250 revolutions per minute. This is precisely the same process as the ventilation of a mine, except that we deal with corn instead of air. The fan is of simple construction, being a hollow wheel with passages radiating from the centre outwards.

The same principle holds good in pumping water.

EXPERIMENTS ON ROTATION.

Some lecture-table experiments will also illustrate this subject.

Conceive that two tubes, each nearly filled with water, are mounted on an axis A B, as shown ; place a cork in one tube and a bullet in the other. Now set the tubes in rapid rotation, and it will be found that the bullet rises, while the cork descends; the air-bubble also descends to the bottom of each tube. The ordinary laws of gravity are reversed in this artificial state of things. The reason is, that the heaviest body exerts the most powerful tendency to pursue its path in a straight line. The water is heavier than the cork, and pushes it back ; the water is lighter than the bullet, and is pushed back by it : so also the water forces down the lighter air.

The bullet is shown oval in form ; a round bullet is apparently distorted into that shape when placed inside a cylindrical glass tube. The tube acts as a lens.

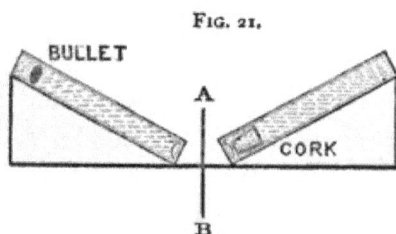

FIG. 21.   FIG. 22.

Again, take a bottle shaped somewhat as in the sketch, pour into it some water and some mercury, and put also into it a white marble and a cork ; then set the bottle in rapid rotation upon a vertical axis. The mercury exerts the greatest power in this attempt to go forward in a straight line, and arranges itself, under the constraint of the vessel, in a horizontal band ; next comes the marble, which describes a circle, looking like a shadowy white ring resting inside the mercury ; then the water hollows itself out in the form of a cup, and the cork appears to be in one sense heavier than the mercury, although it is seen floating in the centre of the hollow formed by the water, where there is a quiet space not affected by rotation.

In the same vessel we observe that the ordinary laws of gravity are both contradicted and confirmed.

SIEMENS' STEAM JET.

The effect produced by a jet of steam in setting air in motion and in creating a partial vacuum has long been noticed. One most useful application occurs in the steam blast which beats out in puffs from the chimney of a loco-motive, and sustains a rush of air through the tubes of the boiler.

Mr. Siemens has recently applied a steam jet for exhaust-ing air, and he has found it essential to make the velocity of the air, just as it comes in contact with the steam, as nearly as possible equal to that of the steam. It was a knowledge of principles which led to this conclusion, and the reason is the same as that for causing the velocity of corn when it falls upon a carrying-band to be equal to that of the band. The corn would, as already explained, have been left behind if it had not this velocity, and so also the air would be left behind by the carrying steam, and would be whirled into eddies with a loss of force, if it had not sufficient velocity at the instant of coming into contact with the steam.

The first law of motion applies equally in both cases, and the requisite velocity is obtained by contracting the air pas-sage up to the point where the jet of steam is in action.

So also, common sense would tell us that the effect will be heightened by increasing the amount of surface where the steam and air are in contact. This is done by pouring out the steam in an annular jet between two concentric annular passages for the air.

To show that the steam jet is a practical contrivance, it may suffice to say that it has been found competent to work the pneumatic despatch tubes used for sending telegrams. The stations from Telegraph Street to Charing Cross are connected by a continuous line of iron pipe three inches in diameter, and forming a complete circuit of two parallel tubes nearly four miles in length. These tubes pass round the corners of streets, dip under buildings, and rise or fall, as may be necessary, but they keep up a continuous communication to Charing Cross and back again. The written messages are placed in light cases covered with drugget, and are blown through the tubes by a steam-engine which drives the air in at one end and pulls it out at the other. As an experiment on the power of the jet, the whole length of four miles of tubing has been worked by three steam-jet exhausters, and the carrier cases have been propelled through the tubes at the rate of fourteen miles an hour without any assistance from the engine.

The student has now learnt to give inertia to water and air as well as to solid bodies, and in doing this, he has made an important step in mechanics. He can see a grain of wheat, but he cannot distinguish a single particle of air, yet the same mechanical laws apply in both cases, and thus we reason from the large and visible to the minute and imperceptible, until finally, we become so far educated as to apply our knowledge of mechanics in explaining the subtle relations concerned in the motion of those invisible atoms which transmit heat and produce effects which we say are caused by force.

We pass on to some examples on the motion of falling bodies.

Ex. 1. A body falls freely from rest for six seconds, what is the space described in the last two seconds of its fall? (Science Exam. 1869).

Taking the expression $s = \frac{1}{2}gt^2$ we have

Space described in 6 seconds $= \frac{1}{2}g \times 36$

,, 4 ,, $= \frac{1}{2}g \times 16$

∴ space described in the last two seconds $= \frac{1}{2}g (36-16) = 322$ feet.

Ex. 2. Find the time in which a body falls from rest through 192 yards. (Science Exam. 1871).

Here also $s = \frac{1}{2}gt^2$, ∴ $192 \times 3 = 16 \cdot 1 \times t^2$, putting yards into feet.

∴ $t^2 = \frac{576}{16 \cdot 1} = \frac{576}{16} = 36$ very nearly, ∴ $t = 6$ seconds.

It is usual to take g as 32 when the answer works out easily by doing so, and in the present case we observe that

$192 = 12 \times 16$, ∴ $12 \times 16 \times 3 = 16t^2$, ∴ $t^2 = 36$ and $t = 6$.

Ex. 3. A stone is thrown upwards with a velocity 64 feet per second, find when it is 48 feet above the ground.

If no force acted the stone would describe $64 t$ feet in t seconds; also the pull of the earth causes it to fall from rest through $16 t^2$ feet in t seconds. We conclude that each of these movements may occur separately or together, and that neither will influence the other, because the second law of motion tells us that although the stone is moving upwards at first, it commences to accept the motion of a body falling freely from rest at the instant when it is liberated from the hand.

Hence $48 = 64t - 16t^2$, ∴ $t^2 - 4t = -3$, and $t = 1$ or 3.

The double answer is easily explained. The stone rises 48 feet in 1 second, but it goes higher, stops, and comes down again, and is 48 feet above the ground, after an interval of 3 seconds.

Ex. 4. To find how high the stone rises.

The stone must rise to the height from which it would have to fall in order to acquire the velocity 64.

Hence, taking the formula $v^2 = 2gs$, we have $64^2 = 2 \times 32 \times s$,

∴ $64 = s$, or the stone will rise through 64 feet.

Ex. 5. To find the time occupied in rising through this height.

Here $s = \frac{1}{2}gt^2$, ∴ $64 = \frac{1}{2} \times 32t^2 = 16t^2$, ∴ $t = \pm 2$.

The double sign is again significant, the stone is 2 seconds in rising to the height, and 2 more in dropping back to the point from which it was thrown upwards.

A question suggested by dropping a stone into Carisbrook well may be taken to illustrate the nature of both uniform and accelerated motion. The stone falls with an increasing velocity till it strikes the water, but the sound of the splash rises uniformly.

Ex. 6. A person drops a stone into a well, and after 3 seconds hears it strike the water. Find the depth to the surface of the water, the velocity of sound being 1127 feet per second.

Let x be the time in seconds occupied by the stone in falling, then the depth of the well $= 16 \cdot 1 x^2$ feet.

But $3 - x$ is the time during which the sound rises uniformly, $\therefore$ the depth of the well $= 1127 (3 - x)$ feet.

Hence $16 \cdot 1 x^2 = 1127 (3 - x)$, $\therefore \dfrac{x^2}{10} = 7 (3 - x) = 21 - 7 x$.

$$x^2 + 70x + 1225 = 1225 + 210 = 1435$$
$$x = 2 \cdot 88 \text{ seconds,}$$

$\therefore$ the depth of the well $= 1127 (3 - 2 \cdot 88) = 1127 \times \cdot 12 = 135 \cdot 24$ feet.

In the text we have proved the formula $s = \frac{1}{2} g t^2$; this proof is general, and will apply to the action of any constant force, so that we may interpret it thus :

space from rest $= \frac{1}{2}$ (vel. generated in $1''$) (time)^2.

Or putting it into symbols, let f be the velocity generated in 1 second by a given uniform force; s the space described from rest in t seconds,

then $\quad s = \frac{1}{2} f t^2$.

Ex. 7. A mass of 500 lbs. is acted on by a force of 125 *absolute units*, what space will it describe from rest in 8 seconds?

In absolute measure a force 1 generates in 1 second a vel. 1 foot per second in a mass of 1 lb., therefore a force of 500 would generate in 1 second a vel. 1 in a mass of 500 lbs., and a force of 125 would generate in 1 second a vel. $\frac{1}{4}$ in a mass of 500 lbs.

Hence, space required $= \frac{1}{2} \times$ (vel. generated in $1''$) (time)^2

$$= \frac{1}{2} \times \frac{1}{4} \times (8)^2 = \frac{1}{8} \times 64 = 8 \text{ feet.}$$

As an example in *gravitation units*, take the following problem.

Ex. 8. A weight of 8 lbs. (called P), is placed on a smooth horizontal table, which does not resist the motion, and is attached by a string to a weight of 12 lbs. (called Q) hanging over the table. Find the tension of the string, the velocity generated in 1 second, and the space described from rest in 2 seconds.

As regards the tension of the string, it is evident that the strain is lessened by the fact that the weight of 8 lbs. yields to the pull on it.

FIG. 23.

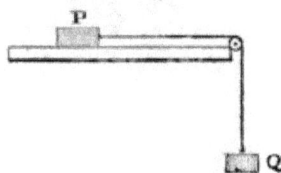

Let f be the velocity generated in one second in P,

T the tension of the string in pounds.

Since T acting on P generates a velocity f, we have

$$\frac{T}{8} = \frac{\text{mass} \times f}{\text{mass} \times g} = \frac{f}{g}.$$

Also $(12 - \text{T})$ acting on Q generates the same velocity,

$$\therefore \frac{12 - \text{T}}{12} = \frac{\text{mass} \times f}{\text{mass} \times g} = \frac{f}{g}.$$

Hence $\quad \dfrac{\text{T}}{8} = \dfrac{12 - \text{T}}{12},$

$$\therefore \quad 3\,\text{T} = 24 - 2\text{T}, \text{ and } \text{T} = 4\tfrac{4}{5} \text{ lb.}$$

Also $\quad \dfrac{f}{32} = \dfrac{\text{T}}{8} = \dfrac{24}{8 \times 5}, \quad \therefore \quad f = \tfrac{96}{5} = 19{\cdot}2.$

Also space described from rest by either P or Q in 2 seconds

$$= \tfrac{1}{2} f \times (2)^2 = 2f = 38{\cdot}4 \text{ feet.}$$

Ex. 9. A body whose mass is 8 lbs. is known to be under the action of a single constant force. It moves from rest, and describes $\frac{5}{2}$ ft. in the first second, what is the magnitude of the force?

(Science Exam. 1871.)

As before, let **F** be the force which generates a velocity 5 in 1 second in a body weighing 8 lbs.

$$\therefore \frac{\text{F}}{8} = \frac{\text{mass} \times 5}{\text{mass} \times 32 \cdot 2} = \frac{5}{32 \cdot 2}, \quad \therefore \text{F} = \frac{5 \times 8}{32 \cdot 2} = \frac{40}{32 \cdot 2} = 1 \cdot 24 \text{ lb.}$$

Ex. 10. A problem in this form appears to be of little practical use, but the knowledge we gain by solving it may be valuable.

Thus, in the *Allen steam-engine*, of which there is an example at the works of Sir J. Whitworth & Co., **the crank** shaft makes 200 revolutions per minute. **The piston and the** parts connected with it weigh **470** lbs. This **mass comes to rest** at the end of one stroke, and must be started **again on its return** ; it is easy to see that the piston may either be **dragged on by its** connection with the rotating shaft and flywheel, or that **it** may be moved by the pressure of the entering steam. The problem now is, *to* find approximately what must be the pressure of the steam in order that the piston may begin to move without straining the crank pin.

Here **the** crank is 12 inches in length, and the number of revolutions is 200 per minute, it follows therefore that the piston moves through ·000152 ft. while the crank describes an arc of 1°, i.e., in $\frac{1}{1200}$th of a second. What then is **the force** which is competent to move 470 lbs. from rest through ·000152 ft. in $\frac{1}{1200}$ seconds?

Let P be the required force in pounds, f the velocity which it generates in 1 second in a body weighing **470** lbs.

$$\therefore \quad \frac{\text{P}}{470} = \frac{f}{g} = \frac{f}{32 \cdot 2}$$

But space from rest $= \frac{1}{2} f.(\text{time})^2$, $\therefore \cdot 000152 = \frac{f}{2} \times \left(\frac{1}{1200}\right)^2$

$$\therefore f = \cdot 000152 \times 1440000 \times 2,$$

$$\frac{f}{32 \cdot 2} = \frac{1 \cdot 52 \times 144}{16 \cdot 1} = 13 \cdot 6,$$

$$\therefore \text{P} = 470 \times 13 \cdot 6 \text{ lbs.} = 6392 \text{ lbs.}$$

The area of the piston is 113 square inches, and hence the steam should be admitted at a pressure of 57 lbs. per square inch nearly.

Ex. 11. A body is thrown upward with a velocity of 96 feet per second. After how many seconds will it be moving *downward* with a velocity of 40 feet per second? Take $g = 32$. (Science Exam. 1872.)
The answer is $4\frac{1}{4}$ seconds.

Ex. 12. A moving body is observed to increase its velocity by a velocity of 8 feet per second in every second. How far would it move from rest in 5 seconds? (Science Exam. 1872.)
Answer. 100 feet.

Ex. 13. A body known to be acted on by a constant force moves from rest, and describes 36 feet in the first 3 seconds. With what velocity will it be moving at the end of the sixth second?
Answer. 48 feet per second. (Science Exam. 1871.)

Ex. 14. A body falls freely from rest through 160 feet. How long will it take to fall through the next 80 feet? Take $g = 32$.
Answer. $\frac{3}{4}$ths of a second, nearly. (Science Exam. 1871.)

Ex. 15. A body weighing 50 lbs. is acted on by a constant force which acts for 5 seconds and then ceases to act: the body moves through 60 feet in the next 2 seconds. Express the force in absolute units.
The answer is 300. (Science Exam. 1870.)

Ex. 16. A body acted on by a uniform force is found to be moving at the end of the first minute from rest with a velocity which would carry it through 10 miles in the next hour. Show that the velocity generated in 1 second by this force : $g :: 1 : 131$ nearly.

Ex. 17. A body falling freely is observed to describe $112 \cdot 7$ feet in a certain second: how long previously to this has it been falling?
Answer. 3 seconds.

There are many other matters for illustration connected with this introductory chapter, but we prefer not to treat of them at present. The text is sufficiently clear as far as it goes, and it will serve the purpose of an introduction to the more detailed information which will be given hereafter.

We pass on to discuss the two leading principles which govern the laws of equilibrium of bodies under the action of forces, viz., the *principle of the lever*, and the *principle of the parallelogram of forces*.

CHAPTER I.

THE PARALLELOGRAM OF FORCES AND THE LEVER.

ART. 1.—It has been stated that force is any cause which moves or tends to move matter, and we have now to suppose that a **very minute** particle of matter **at A** is acted on at the same instant **by two forces,** P and Q. Unless **these forces are equal, and** act in opposite **directions in the** same straight line, **the** molecule must move in some determinate direction with a definite velocity. **But a single** force is competent to move a molecule in **a** definite direction with an assigned velocity. Hence, there is some single force R, intermediate to P and Q, which, when acting on the molecule, produces **the same effect as the** combined action of P and Q.

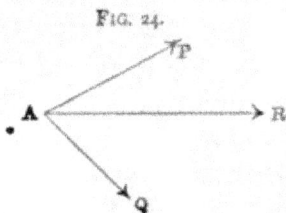

This single **force** R is called the *resultant* **of** P and Q, **and conversely,** P and Q **are** called *components* of R. The **process of** substituting two forces P and Q for the single force R **is called the** *resolution of force,* and the process of finding R from **P and Q** is called the *composition of the forces.* In other words **we say** that R may be *resolved* into P and Q, or that **P and Q** may **be** *compounded* into the single force R.

In flying **a kite,** for example, there are two forces acting externally, viz., **the pressure** of the wind and the pull of the string; a third force is **the weight of** the kite, which must be equal and opposite to the resultant of the two first-named forces. The kite will **remain steadily** supported in the air so long as **this equality maintains.**

If any number of forces acted at the same instant on the point A, there would still be a single force competent to move A as it actually moves. Hence, the forces would have a single resultant, and the process of resolution or composition may be extended to any number of forces.

THE REACTION OF SMOOTH SURFACES.

2. It will now be convenient to explain the meaning of the term reaction, as applied to bodies when at rest under the action of forces.

A particle is free when there is nothing to prevent its motion in any direction, but a particle is constrained when there is some particular direction in which it cannot move. Thus, a body placed upon a horizontal table is constrained and not free, for it cannot penetrate below the table. A body moving in a curved tube is constrained, and so is a body suspended by a string.

Whenever this constraint occurs, there is a resistance brought into play which is called *reaction*. When a body rests on a table, its weight presses on the table and produces an *action*, the table supports this pressure and exerts a *reaction*, which, by Newton's 3rd law, must be equal and opposite to the action.

The sides of the tube react on the body moving within it. The weight hanging on a string produces a stretching force or tension, and this tension is an upward pull or reaction equal to the action of the weight sustained.

The next point to be considered is the direction of this reaction, and here experiment leads us to conclude that if the surface of constraint be absolutely smooth, the reaction must be always perpendicular to the surface.

What is meant by a smooth surface ? An optician would say that it is a surface which does not scatter or throw off irregularly any part of a beam of light, and that if there were such a thing as a smooth surface we could not see it, because none of the light falling on it would be scattered.

We should, in point of fact, only see the objects which were reflected in it.

In natural objects we **never find this** ideal perfection, and the test of smoothness is **simply** the absence of any resistance offered by the surface **to motion along itself.** Thus, ice is called smooth because **a body slides along it so** easily. **The** most level and true **plane** that **can be** made of metal **is not smooth in** the strict sense of the word, for resistance to motion along its surface is felt at once.

It suffices to point out that an ideal smooth surface offers no resistance to motion in any direction except in a line perpendicular to itself. We always assume, in problems **relating** to bodies resting on smooth surfaces, **or** relating **to** bodies having smooth surfaces and resting on supports, that the reaction is **in every** case perpendicular to the surface.

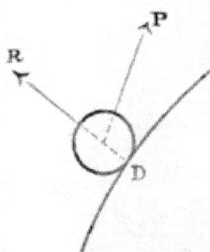

FIG. 25.

For example, if a smooth sphere, held up by a string P, rests on a smooth surface at the point D, the reaction R of the surface will be at right angles **to** the direction of the common surfaces **at D,** and **the** pressure exerted on the supporting surface at D will be equal to the force R.

PRINCIPLE OF THE CONCURRENCE OF THREE BALANCING FORCES.

3. Before discussing the method of finding the resultant of two **or more** forces acting on a point, we shall explain a principle known as that of the concurrence of three balancing forces, which **may be** thus stated.

When three forces act on a body and keep it at rest they must either meet in one point or not meet at all.

If they do not **meet they** are parallel, and that case will be examined presently. **If** they **do meet** they must all come together in one point. For two at least of the forces meet in one point and have a single resultant ; this resultant balances the **third force** ; but **two** forces which balance act in the same straight line and in opposite directions, therefore the third force must pass through the point of intersection of the other two forces.

In the following examples of this principle, we shall assume that the weight of a uniform rod acts as if it were collected in the middle point of a rod. The proof will be given very shortly.

Ex. I. A uniform rod rests in a hemispherical cup as shown, find its length when it rests at 30° to the horizon.

FIG. 26.

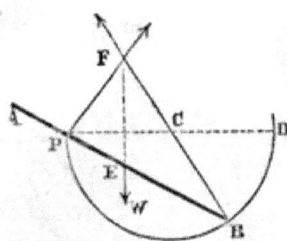

Let E be the centre of the rod, then there are only 3 forces acting, viz., the reaction at P, the reaction at B, and the weight of the rod; and it follows that these 3 forces must meet in a point F.

Then F P C = 60°, since the rod is inclined at 30° to the horizon.

But F C P = B C D = 2 C P B = 60°,

∴ F P C = F C P = 60°,

∴ P F C is also 60°,

∴ F C P is an equilateral triangle.

Let the radius of the hemisphere = a, and length of rod = $2x$.

$$\therefore \frac{x}{FB} = \frac{\sin EFC}{\sin FEB} = \frac{\sin 30}{\sin (90 + 30)} = \frac{\sin 30}{\cos 30} = \frac{1}{\sqrt{3}}$$

$$\therefore \frac{x}{2a} = \frac{1}{\sqrt{3}}, \qquad \therefore x = \frac{2a}{\sqrt{3}}$$

Ex. 2. A uniform beam A B hinged at A, and weighing 100 lbs., rests in a horizontal position under the pull of a weight W attached to a string making an angle of 30° with the beam. Find W and the pressure on the hinge.

The forces acting are this pressure, which call P, the force W, and

FIG. 27.

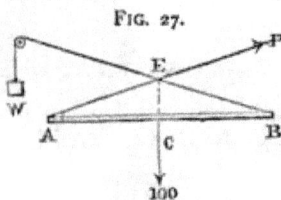

the weight of the beam, viz., 100 lbs. Now the directions of the tension at B and the weight of the beam meet in E, therefore P also passes through E.

Also P, W, and 100 make equal angles with each other, and no reason can be assigned for making any one force greater than either of the other two forces.

∴ they must be equal to each other,

∴ P = W = 100.

Ex. 3. A uniform rod B E rests on a prop A, and also against a smooth vertical wall B D.   Find the length of the rod when it makes an angle θ with the wall, the distance from the prop to the wall being given.

FIG. 28.

Here R′, the reaction at B, is a horizontal force, and w, the weight of the rod, acting through its middle point G, is a vertical force.

These two forces meet in F, and the reaction R at the prop A must pass through the same point.

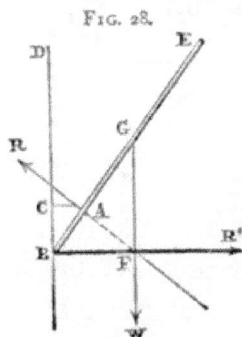

Now $A C = A B \sin \theta$

and $A B = B F \cos A B F$

$\quad = B F \sin \theta,$

$\therefore \quad A C = B F \sin^2\theta \quad = B G \sin^2\theta.\ \sin \theta \quad = B G.\ \sin^3\theta.$

$\therefore \quad B G = \dfrac{A C}{\sin^3\theta},$ and $B E = \dfrac{2 A C}{\sin^3\theta}.$

Ex. 4. Let θ be $\frac{1}{3}$ of a right angle $\therefore \sin \theta = \frac{1}{2}$, and $B E = 16.\ A C.$

Ex. 5. Draw a vertical line A B, and a line A C making an angle of 30° with it ; place an equilateral triangle P Q R, of weight w, between these lines, with one angle on A B and the other two on A C.   Assume that the surfaces are smooth, and find the pressures on A B, A C.

The forces make angles of 90, 120, 150 with each other, whence the pressures are $w \sqrt{3}$, and 2 w.   (Science Exam. 1870.)

THE PRINCIPLE OF THE PARALLELOGRAM OF FORCES.

4. We pass on to a mechanical principle of the highest value called the *parallelogram of forces*, which may be thus enunciated.

If two straight lines drawn from a point represent in magnitude and direction any two forces acting on that point, and if the parallelogram on these lines be completed, the resultant of the two forces will be represented in direction and magnitude by that diagonal of the parallelogram which passes through the point where the forces act.

The reasoning is divided into two distinct steps, (1) we find the *direction* of the resultant, and (2) we find its *magnitude*.

1. To show that this general proposition is true *so far as regards the direction of the resultant.*

F

Conceive that two equal forces P, P, represented by the
straight lines A B, A C, act upon the material
particle A.* The direction of their resultant R
must bisect the angle B A C, for no reason
can be adduced for its inclining towards
one of the forces, as A B, which would not
equally apply to make it incline towards the
other, as A C; that is, its direction must lie exactly half way
between A B and A C.

FIG. 29.

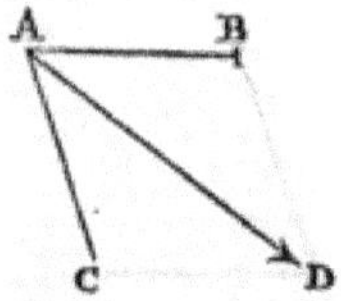

If now we complete the parallelogram A B C D, as in Fig. 30,
and join A D, the diagonal A D bisects
the angle B A C, and therefore A D
determines the direction of the resultant of the forces P, P. Hence the
proposition is true, so far as regards
the direction of the resultant, in the
case when the forces are equal.

FIG. 30.

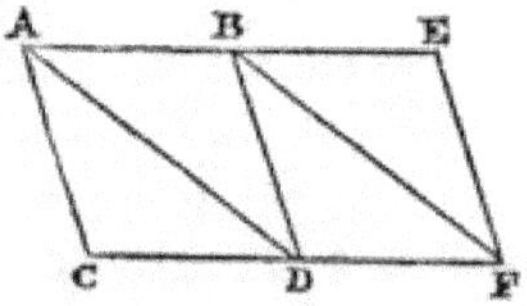

Next conceive that the point A is acted on by a force P
in A C, and by two forces P, P, in direction A E.

Make A C=P, A B=P, B E=P, complete the parallelograms
B C, E D, and join A D, B F. Let the diagram represent a number
of rigid lines immovably fastened together, then the forces
P, and 2 P, acting at A, will have a resultant in some direction as yet unknown.

We argue thus respecting it; if we can transfer the
forces P and 2 P to another point, such as F, for example,
without disturbing the action felt at A, we shall have a right
to conclude that the resultant of P and 2 P lies in the
direction A F. *The principle of the transmission of force†*
justifies this conclusion, for it is not possible to transfer a
force, whether compound or single, to any point not situated
in the line of its action. In order to follow out this train of
thought we regard the point A as acted on directly by the
forces P, P, in A B, A C, and conceive that the second force P

* See page 2 of the Introduction.
† See page 3 of the Introduction.

is applied at B and transmits its action on to A by means of the rigid line B A.

Now P in A B, and P in A C, have a resultant bisecting the angle B A C, and therefore acting in A D. Transfer this resultant to D, and then resolve it back again into its components, viz., P in B D and P in D F; transfer the first named component to B by the rigid line D B, and the second component to F by the rigid line D F. Although the points of action of the forces are changed, the pull at A is precisely the same as at first.

Now the force P at B, in B D, will compound with the force P in B E, and these two forces P, P acting at B will have a resultant in direction B F. Transfer this resultant to F, and break it up into its components P, P. We shall then have P at F in direction E F, P at F in direction D F, as well as a third force P at F, in direction D F, which was carried there previously.

On the whole there are now P and 2 P at F acting in directions parallel to P and 2 P at A. That is, the forces P, 2 P acting at A have been shifted to F without disturbing the action at A. We conclude, therefore, that the direction of the resultant of P and 2 P at A lies in the line A F, which is the diagonal of the parallelogram on the sides A E, A C. If, therefore, the proposition, so far as regards the direction of the resultant, be true for the forces P and P, it is also true for the forces P and P+P, or P and 2 P. In like manner, if it be true for A C and A B or for P and P, and also for A C and A E or for P and 2 P, it may be shown to be true for A C and A B + A E or P and 2 P + P, i.e. for P and 3 P. By extending this reasoning, the general proposition, restricted to the direction of the resultant, is true for P and m P, and finally for n P and m P; that is, for all commensurable forces.

It may also be proved to be true for *incommensurable* forces, that is, forces represented by the quantities such as $5 \sqrt{2}$, $\sqrt{3}$, and the like. We pass by any such refinements of reasoning, and the proof must suffice as it stands.

2. To find the *magnitude* of the resultant. Let A B, A C, represent, in magnitude and direction, two forces P and Q acting A; complete the parallelogram A B C D, join A D; then A D gives the direction of the resultant of P and Q. Let R be the magnitude of this resultant, produce D A to K making A K = R, complete the parallelogram K C, and join A H. Since the forces A K, A B, A C, acting A are in equilibrium, any one may be regarded as equal and opposite to the resultant of the other two, that is A B is equal and opposite to the resultant of A K, A C. But if this be so, A B is in the direction of A H, the diagonal of the parallelogram K C;

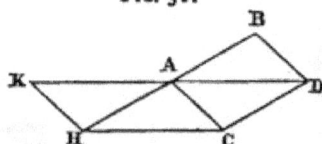

Fig. 31.

∴ H A B is a straight line.

∴ H A is parallel to C D, and H A D C is a parallelogram.

∴ H C is equal to A D.

But H C = K A ∴ K A = A D.

But K A = R ∴ A D = R.

Therefore the diagonal of the parallelogram C B represents the resultant of P and Q in magnitude as well as in direction, and the general principle is established.

5. The relations between R, P, Q in the general proposition are found at once by trigonometry.

Let B A C = a, B A D = θ.

Then A D^2 = A C^2 + C D^2 − 2 A C . C D . cos A C D

or $\quad$ R^2 = P^2 + Q^2 + 2 P Q cos a . . . . . . (1)

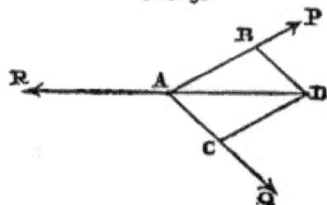

Fig. 32.

Also $\dfrac{R}{P} = \dfrac{A D}{A B} = \dfrac{\sin a}{\sin (a - \theta)}$

$\dfrac{R}{Q} = \dfrac{A D}{A C} = \dfrac{\sin a}{\sin \theta}$

These equations give R : P : Q

$= \sin a : \sin (a - \theta) : \sin \theta$

$= \sin$ P A Q : $\sin$ R A Q : $\sin$ R A P.

Cor. 1. The resultant of two *equal* forces may be found by referring to equation (1), and making $Q = P$.

It follows that

$$R^2 = P^2 + P^2 + 2 P^2 \cos a$$
$$= 2 P^2 (1 + \cos a)$$
$$= 4 P^2 \cos^2 \frac{a}{2}$$
$$\therefore \quad R = 2 P \cos \frac{a}{2}.$$

The same thing appears from the figure, for when the sides of a parallelogram are all **equal**, the diagonals bisect each other **at right** angles.

Let E be this **point of** intersection; then

$$R = A D = 2 A E = 2 A B \cos B A D = 2 P \cos \frac{a}{2}.$$

Cor. 2. The greatest value of the resultant is $P + Q$, and the least value is $P - Q$; also the resultant increases as the angle between the forces diminishes.

RECTANGULAR COMPONENTS OF A FORCE.

6. In applying this principle it may be convenient to commence with the case where B A C is a right angle. Let the forces X and Y, acting at a right angle on the point A, be represented in magnitude and direction by the lines A B, A C.

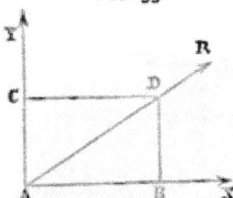

FIG. 33.

Then R, the resultant of X and Y, will be represented in magnitude and direction by A D, the diagonal of the parallelogram A B D C.

Let B A D $= a$.

Then $AB = AD \cos a$, or $X = R \cos a$ . . . . (1)

$\qquad AC = AD \sin a$, or $Y = R \sin a$ . . . . (2)

$\qquad \therefore X^2 + Y^2 = R^2 (\cos^2 a + \sin^2 a) = R^2$

$\qquad\qquad \therefore R = \pm \sqrt{X^2 + Y^2}.$ . . . . . (3)

Equations (1) and (2) give the components of R in directions A B, A C respectively, and (3) gives R in terms of X and Y.

From this we conclude that a force R may be resolved in and perpendicular to a line making an angle a with its direction by multiplying it by cos a and sin a respectively.

Also $$\frac{R \sin a}{R \cos a} = \frac{Y}{X} \qquad \therefore \tan a = \frac{Y}{X}.$$

Ex. 1. Resolve the force 10 into two forces at right angles to each other, one of the forces making an angle of 20° with the force 10.

Here $X = R \cos 20 = 10 \times \cdot 940 = 9 \cdot 4$

  $Y = R \sin 20 = 10 \times \cdot 342 = 3 \cdot 42.$

Ex. 2. Two forces, each of 100 lbs., make angles of 30° and 45° with a horizontal plane, and act in directions opposed to each other at a point in the plane. Find the effect along the plane.

Here $X = 100 \cos 30 - 100 \cos 45$

   $= 100 \times \cdot 866 - 100 \times \cdot 707$

   $= 86 \cdot 6 - 70 \cdot 7 = 15 \cdot 9$

RECTANGULAR COMPONENTS OF SEVERAL FORCES.

7. This principle of the resolution of a force into two components at right angles to each other, may be extended to any number of forces acting in one plane upon a point.

FIG. 34.

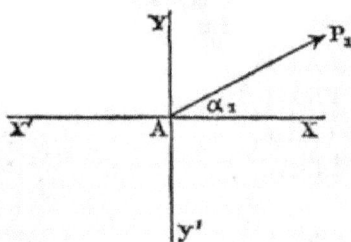

Let X A X′, Y A Y′, be two straight lines at right angles to each other, meeting in the point A. P_1 a force, acting at A and making an angle a_1 with A X; also let other like forces P_2 P_3 &c., acting at A, make angles a_2 a_3 &c., with A X.

Then $P_1 \cos a_1 + P_2 \cos a_2 + P_3 \cos a_3 +$ &c., is the sum of the components of these forces along X A X′, and $P_1 \sin a_1 + P_2 \sin a_2 + P_3 \sin a_3 +$ &c. is the sum of the components of these forces along Y A Y′.

Let R be the resultant of all these forces, θ the angle which it makes with X A X′,

then $R \cos \theta = P_1 \cos a_1 + P_2 \cos a_2 +$ &c. $= X$

   $R \sin \theta = P_1 \sin a_1 + P_2 \sin a_2 +$ &c. $= Y$

 $\therefore R^2 = X^2 + Y^2,$ and $\tan \theta = \frac{Y}{X},$

Cor. If R = 0, then $x^2 + y^2 = 0$,

$$\therefore x = 0, \text{ and } y = 0,$$

that is the components in directions $x \, A \, x'$, $y \, A \, y'$, are separately in equilibrium. We shall make great use of this simple statement.

Ex. 1. Three forces of 27, 52, 49 lbs., act at the point o in directions o A, o B, o C, such that A O B = 32°, B O C = 26°. Find their resultant.

Here $R \cos \theta = 27 + 52 \cos 32 + 49 \cos 58 = 97 \cdot 0645$

  $R \sin \theta = 52 \sin 32 + 49 \sin 58 = 69 \cdot 1102$

$\therefore$ $R^2 = 14197 \cdot 7, \ R = 119 \cdot 4$.

Ex. 2. Three forces, 4, 5, 6, act on a point at 120° to each other; find their resultant, and the angle at which it is inclined to the force 4.

Here $x = 4 - 5 \cos 60 - 6 \cos 60 = -\frac{3}{2}$

  $y = 5 \sin 60 - 6 \sin 60 = -\frac{1}{2} \sqrt{3}$

$\therefore$ $R^2 = \frac{9}{4} + \frac{3}{4} = 3$ and $R = \sqrt{3}$.

Let θ be the required angle, $\therefore \tan \theta = \frac{y}{x} = \sqrt{\frac{1}{3}}$ $\therefore \theta = 30$ or 210.

It is clear that 210° is the angle required by the circumstances of the problem.

The concluding proportion between R, P, Q, (*See Art.* 5) is often stated as follows :—

When three forces act on a point and keep it at rest, each one of them is proportional to the sine of the angle between the other two.

It also takes another form, which was an early invention in mechanics, and is known as the *triangle of forces.*

THE TRIANGLE AND POLYGON OF FORCES.

8. *If three forces acting on a point be represented in magnitude and direction by the sides of a triangle taken in order, they will be in equilibrium.*

Let A B D be a triangle, whose sides, A B, B D, D A, taken in order, represent in magnitude and direction, the forces A B, A C, D A, acting at the point A. Complete the parallelogram A B C D. Then the forces A B, A C, have a

Fig. 35.

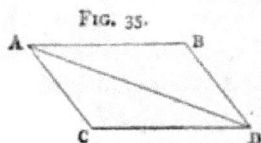

resultant A D, that is, the forces A B, A C, D A, are equivalent
to A D, D A, and, therefore, balance each other.

It follows that the forces represented by A B, B D, D A, would
be in equilibrium if they were applied directly at the point A.

The converse is also true, viz., that if the forces balance,
and the triangle be constructed, its sides will be proportional
to the magnitudes of the forces.

This statement governs the application of the proposition
in practice. A triangle is drawn whose sides are parallel to
the forces, the sides of this triangle are measured, or found
by calculation, and give the respective magnitudes of the

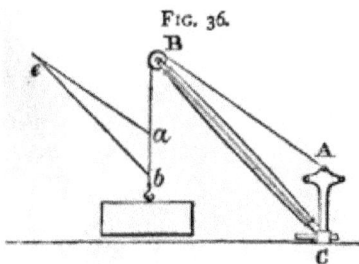

FIG. 36.

forces. There are numerous
examples in the text-book on
the strength of materials.
Take the case of a crane.
Let a weight of four tons hang
on the chain of a crane from
the end c, take ab = four units,
draw ac, bc, parallel to the
tie-rod A B and the jib B C respectively. The lines ac, bc,
may be found either by measurement or calculation, and
they will represent, on the same scale as ab, the pull on the
tie-rod and the compressing thrust on the jib respectively.

Ex. 1. Three forces of 8, 10, 12 units respectively keep a point
in equilibrium; determine by construction how they act.

Make a triangle whose sides are 8, 10, 12 respectively, and the
directions of the forces are parallel to the sides of this triangle.

(Science Exam. 1872.)

Ex. 2. A B C is an equilateral triangle. Forces of 10, 10, 15 lbs. act
on a point in directions parallel to A B, B C, C A, respectively; find their
resultant. *Ans.* The resultant is 5 lbs.

Ex. 3. A rod A C, without weight and hinged at C, supports a weight
of 100 lbs. hung at A, and is kept in position by a horizontal tie-rod A B.
If B A C is 30°, find the tension of the tie-rod and the thrust on A C.

Ans. Tension = 100 $\sqrt{3}$ lbs.   Thrust = 200 lbs.

Ex. 4. Find the resultant of forces of 10, 20 lbs. acting on a point at
an angle of 60°.   *Ans.* 26·4 lbs.   (Science Exam. 1871.)

Ex. 5. Draw the two straight lines A B, A C, at right angles to each other, and bisect the angle B A C by the line A D. A force of 100 lbs. acts in A D, and is balanced by two forces acting in B A, C A. Find these forces. (Science Exam. 1870.)

Ex. 6. **Two** equal forces act at any **point in** the circumference **of a** circle, **and** their directions always pass through fixed **points A and B,** in that circumference. Show that their resultant also **passes through a fixed point** and find it.

Ex. 7. The resultant of two forces is 10 lbs., one of the forces is 8 lbs., and the other is inclined at 36° to the resultant. Find it. *Ans.* 2·66.

There are two solutions, this being the *ambiguous case* in the solution of a triangle. The second answer is 13·52. Two triangles can be drawn having one side equal to 10, another equal to 8, and the angle 36° opposite to the smaller side, viz. 8.

Ex. 8. A point is kept at rest by forces of 6, 8, 11 lbs. Find the angle between the forces 6 and 8. *Ans.* 77° 21' 52".

Ex. 9. Two forces of 123 and 74 lbs. act at an angle of 65°. Find approximately the inclination of each force to the resultant.

Ans. 41° 28' and 23° 32'.

Ex. 10. Find the resultant of two equal forces, each of 20 lbs., acting at an angle of 35°. *Ans.* 38·15.

Ex. 11. A uniform beam weighing 100 lbs. hangs by two cords, which make angles of 108° and 95° with **the beam.** Find the tension of each cord. *Ans.* 52·2 and 49·7.

Ex. 12. Resolve a force of 20 lbs. into two others, whose sum is 22 lbs. and which contain **an angle of 60°.** *Ans.* 17·1, 4·9.

9. The proposition of the triangle of forces is easily extended into that of the polygon of forces, viz.—*If any number of forces acting on a point be represented in magnitude and direction by the sides of a polygon, taken in order, they will be in equilibrium.*

In the polygon A B C D E, we know that A B, B C are equivalent to a force represented by A C, that A C, C D are equivalent to A D, and that A D, D E are equivalent to A E.

∴ A B, B C, C D, D E, E A are equivalent to A E, E A, and will balance each other. Hence the proposition is true.

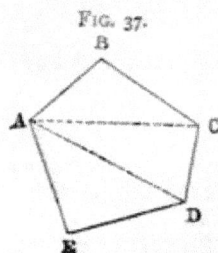
FIG. 37.

THE RESULTANT OF TWO PARALLEL FORCES.

10. Let P and Q represent two parallel forces acting on a rigid body at the points A, B, and in like directions.

Join A B, and apply at A, B, two forces S, S, acting in opposite directions in the line A B ; this will not affect the equilibrium, since these supplementary forces already balance.

Let the forces S and P at A, have a resultant T, and the forces S and Q at B, have a resultant V. Let the forces T, V, meet in C, supposed rigidly connected with A B, and resolve them back again into their components (P, S), (Q, S).

FIG. 38.

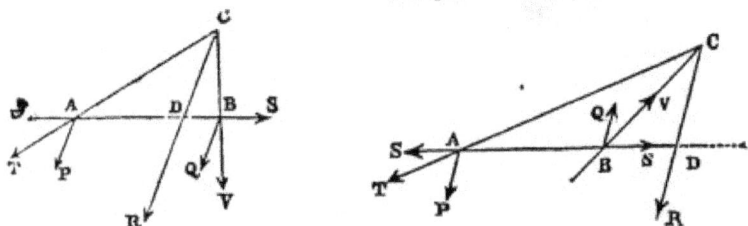

The forces S, S, acting at C, will balance, and may be removed. There will remain P+Q, acting at C, in a direction C D parallel to A P or B Q, and this is the resultant of P and Q, acting at A and B respectively.

Let R represent the resultant of P and Q, then

$$R = P + Q.$$

To find where C R cuts A B, let A B $= a$, A D $= x$.

Then

$$\frac{P}{S} = \frac{CD}{AD}, \quad \frac{Q}{S} = \frac{CD}{DB},$$

$$\therefore \frac{P}{Q} = \frac{DB}{AD} = \frac{a-x}{x} = \frac{a}{x} - 1$$

$$\therefore \frac{a}{x} = \frac{P+Q}{Q},$$

$$\therefore x = \frac{Qa}{P+Q} = DA,$$

and similarly $\quad DB = \frac{Pa}{P+Q}.$

If Q be negative and act in the opposite direction, as in the second figure, we have R equal to the difference of the parallel forces of P and Q as regards magnitude, and having the direction of the greater, and the proof is the same, except that the line C D is shifted to the outside of A B.

$$\text{Also} \quad A D = \frac{Q\,a}{Q - P}, \qquad D B = \frac{P\,a}{P - Q}.$$

This proof may be extended to any number of parallel forces, and we shall now show that their resultant is equal to the algebraical sum of the separate forces.

THE RESULTANT OF **ANY** NUMBER OF PARALLEL FORCES.

11. It has been shown that if two parallel forces P and Q act on a rigid body at the points A,B, in like directions, their resultant is a force P + Q acting at a point C such

that $A C = \dfrac{Q \times A B}{P + Q}.$

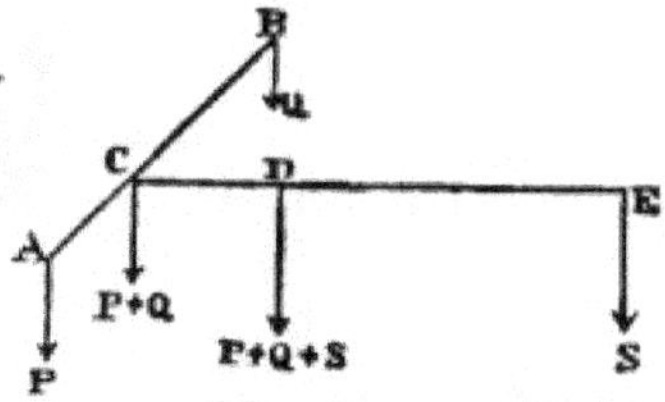

In like manner, the resultant of P + Q at C, and S at E, is P + Q + S acting at a point D, such that

$$C D = \frac{S \times C E}{P + Q + S}.$$

FIG. 39.

By extending this reasoning it is evident that the resultant of any number of parallel forces is equal to the algebraical sum of the several forces.

It is also seen that the resultant of P and Q acts in a line through C, the resultant of P, Q, S, acts through D, and so on. In this way we arrive at the so-called *centre* of all the parallel forces, viz. the point at which their resultant acts.

This process is effective, but it is clumsy, and we shall hereafter give a better method of ascertaining the position of the centre of any number of parallel forces.

THE **CENTRE** OF GRAVITY.

12. It may be well now to introduce a notice of the identity of the centre of parallel forces with that point, known to every one, and called the *centre of gravity* of a body.

If the parallel forces, hitherto considered, be the weights of the respective molecules or particles forming a solid body, it is legitimate to infer that there is a definite line in which the resultant or aggregate of all the weights of the individual particles may be supposed to act. If the body be held in different positions, the lines of direction of the resultant will all intersect in one point. That point is the centre of gravity of the body.

Conceive now that a set of equal and heavy particles are placed at equal distances along a rigid line without weight. Their weights form a system of equal and equidistant parallel forces, and the centre of the line must be the centre of the forces due to the weights of the particles. Also the resultant force is the sum of the weights of the individual particles, and therefore the weight of the series of particles produces the same effect as if it were collected in the centre of parallel forces, or in the centre of gravity. For this reason we always consider that the weight of a uniform straight rod or line of particles, is, so to speak, *collected at its middle point.*

Hereafter we shall discuss the properties of the centre of gravity more fully, but at present we pass on to the examination of another principle, which is that of the lever. This principle might be deduced from the parallelogram of forces, but we prefer to give an independent proof, which is of great interest, as it was invented by Archimedes.

THE PRINCIPLE OF THE LEVER.

13. A *lever* is a rigid bar or rod in which there is a fixed point or axis about which it can freely turn.

The fixed point or axis is called the *fulcrum* of the lever. The fulcrum divides the lever into two parts, called *arms;* when the arms are in the same straight line we have a *straight lever*, in other cases the lever is called a *bent lever.*

The shape of the arms is not material, they may be curved or bent in any way, and in fact any inflexible body which is moveable about an axis constitutes a lever. Thus a wheel is commonly used as a lever, and a system of toothed wheels working together in machinery is merely a combination of levers. We attach the idea of arms to a lever because we are compelled to draw and measure definite lines called arms, in order to estimate the power of the instrument.

FIG. 40.

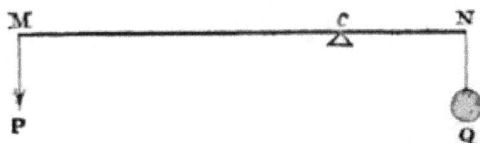

Let M C N represent a rigid horizontal rod without weight moveable in a vertical plane about a fulcrum at C, and let the power P acting vertically at M support the weight Q hung at N. The rule, *called the principle of the lever*, asserts that when there is equilibrium we shall have

$$P \times CM = Q \times CN.$$

14. The proof is based on the observed fact that a uniform cylindrical rod will balance when supported on its middle point.

Let A B represent a uniform heavy cylinder whose weight is P + Q, then A B will balance in a horizontal position on its middle point C.

Divide A B in D, so that the weight of A D shall be equal to P, then the weight of D B will be equal to Q. Bisect A D, D B in M and N respectively.

FIG. 41.

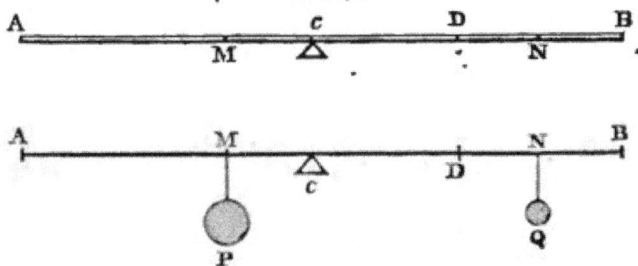

Since the rod A D would balance about M, the weight of A D may be supposed to be collected at M, and similarly the weight of D B may be supposed to be collected at N.

But when these weights are respectively collected at the points M and N, we may suppose them to be hung from those points on a rigid line A B *without weight*. This will not disturb the equilibrium, and the weights P and Q hanging on A B will replace the heavy cylinder, and balance on the point C.

Our object now is to find the relation between P, Q, C M, and N C.

$$\text{Since} \quad \text{C M} = \text{C A} - \text{A M} = \tfrac{1}{2}\,\text{A B} - \tfrac{1}{2}\,\text{A D} = \tfrac{1}{2}\,\text{D B}$$
$$\text{C N} = \text{C B} - \text{B N} = \tfrac{1}{2}\,\text{A B} - \tfrac{1}{2}\,\text{B D} = \tfrac{1}{2}\,\text{A D}$$

$$\therefore \frac{\text{C M}}{\text{C N}} = \frac{\tfrac{1}{2}\,\text{D B}}{\tfrac{1}{2}\,\text{A D}} = \frac{\text{D B}}{\text{A D}} = \frac{\text{Q}}{\text{P}}.$$

$$\therefore \text{P} \times \text{C M} = \text{Q} \times \text{C N}.$$

which proves the proposition.

The principle of the lever is one of those numerous propositions which are true conversely as well as directly.

If the weights balance, the product P × C M is equal to that of Q × C N; and conversely, if P × C M is equal to Q × C N, the weights will balance. It is possible to reason back by the same course reversed.

Ex. 1. Weights of 3 and 8 ounces balance on a straight lever without weight, the longer arm of which is 2 feet. Find the length of the shorter arm.

FIG. 42.

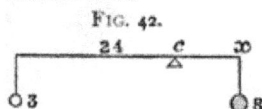

Let $c\,x$ be the shorter arm,

Then $3 \times 24 = c\,x \times 8$,

$\therefore\ c\,x = 9$ inches.

Ex. 2. A rod, whose weight can be neglected, rests on two points 12 inches apart; a weight of 10 lbs. hangs on the rod between the points, and 4 inches from one of them. What is the pressure on each point.

(Science Exam. 1872.)

FIG. 43.

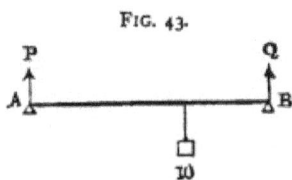

Let P and Q be the pressures at A, B.

Considering B as a fulcrum of the lever A B, we have

$$\text{P} \times 12 = 10 \times 4, \quad \therefore\ \text{P} = \tfrac{10}{3} = 3\tfrac{1}{3}\ \text{lbs}.$$
$$\therefore\ \text{Q} = 10 - \text{P} = 6\tfrac{2}{3};$$

or again, Q × 12 = 10 × 8,

$$\therefore\ \text{Q} = \tfrac{20}{3} = 6\tfrac{2}{3}\ \text{lbs}.$$

Ex. 3. A circular plate of weight, w, is hung at the point A and thrust out of the vertical by a weight P, hanging by a string over its circumference at D. Find the angle which A C makes with the vertical.

Let A C $= a$, draw A E vertical, and meeting D C in E; let E A C $= \theta$. Then we may regard D E C as a lever whose fulcrum is E, therefore P × D E $=$ W × E C.

FIG. 44

$$\text{or} \quad P a (1 - \sin \theta) = W a \sin \theta,$$

$$P - P \sin \theta = W \sin \theta,$$

$$\sin \theta = \frac{P}{W + P}$$

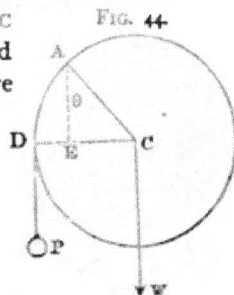

THE MOMENT OF A FORCE.

15. In examining the principle of the lever it is evident that the forces tend to turn the lever round C in opposite directions.

Since P × C M $=$ Q × C N, let C N $= 1$, or let Q act at a constant arm, its effect will therefore remain constant, and Q may be employed as an unit of reference.

Suppose that P $= 10$, C M $= 6$, then Q $= 10 × 6 = 60$. Next let P $= 20$, C M $= 6$, then Q $= 20 × 6 = 120$. In other words, when P is doubled, its turning effect is doubled.

Next let P $= 10$, C M $= 6$, then Q $= 60$ as before ; whereas if P $= 10$, C M $= 3$, we shall have Q $= 30$. Hence we infer that the turning effect of P increases or decreases in the same proportion as C M increases or decreases.

As in page 19 of the introduction, we represent this state of things by asserting that the turning effect of P depends on the product of P and C M, i.e., on P × C M.

It is convenient to give a name to a product which expresses a result of so much importance, and hence it is usual to call P × C M the *moment* of the force P round the fulcrum C. The term moment has an extended meaning in mechanics, and indicates the numerical *measure of the importance of any physical agency.*

Similarly, Q × C N represents the moment of the force Q round the axis through C.

The principle of the lever asserts that these moments are equal when the forces balance.

THE BENT LEVER.

16. If the moment of P represents the power which P exerts to turn the lever C A round a fulcrum at C, then it will be

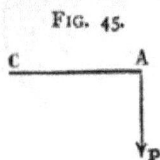
FIG. 45.

competent for us to estimate the power of any other force acting upon any other arm in like manner by its moment, and the direction in which the arm lies will not affect the result. In other words, the turning effect of P on C A is the same whether C A points in one direction or another.

Thus we deduce the condition of equilibrium of a bent lever at once from that of a straight lever. We regard nothing but the equality of moments. All problems upon the bent lever are solved by one uniform method.

For example, let M C N represent a bent lever acted on by forces P and Q, in directions perpendicular to C M, and C N.

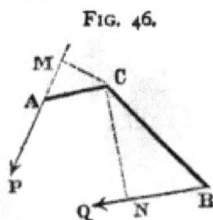
FIG. 46.

The moment of P is P × C M, and that of Q is Q × C N, and the condition of equilibrium is that these moments shall be equal, or that

$$P \times C M = Q \times C N.$$

Next, draw C A, C B, inclined at any angles to M P, N Q, and let A C M, B C N, represent triangles with rigid sides. By the principle of the transmission of force, P and Q may be transferred from M and N, to A and B, respectively; and the condition of equilibrium for the lever A C B will still be, P × CM = Q × C N.

This proposition includes also the case of the straight lever, where the forces act obliquely to the arms, and where the same law of the equality of moments holds good.

17. There is a distinction which has been preserved in books on mechanics, as carefully as if it were of great value, viz., the separation of levers into three kinds or classes ; one, where the fulcrum is between the so-called power and weight, two more where the power and weight are on the same side of the fulcrum. A crowbar is a lever of the first kind ; an oar is a lever of the second kind, the fulcrum being the end of the blade in the water ; and a pair of sheep-shears is a double lever of the third kind. No practical mechanic cares about these distinctions, the turning effect of a force is represented by its moment : that is the lesson to be learnt from the lever, and after this *one* idea is grasped, and the position of the fulcrum is ascertained, the subject-matter is exhausted.

THE PRINCIPLE OF THE LEVER STATED GENERALLY.

18. What has been inferred will be true however we multiply the number of the forces. In order to arrive at the complete turning effect of a number of forces acting in one plane, it suffices to add their separate moments, for in doing so we are only adding numbers of the same kind, which is perfectly admissible.

The principle of the lever may therefore be stated generally as follows.

If any number of forces acting on a rigid body in one plane tend to turn it about a fixed axis, there will be equilibrium when the sum of the moments of the forces acting to turn the body in one direction is equal to the sum of the moments of the forces acting to turn it in the opposite direction.

That point in the axis about which the moments are estimated is often called the *centre of moments.*

THE CONDITIONS OF EQUILIBRIUM OF ANY NUMBER OF FORCES ACTING ON A BODY IN ONE PLANE.

19. Here we arrive at the resting-point to which the previous propositions have all been tending, and when we master these conditions of equilibrium the power of working mechanical problems will be wonderfully increased.

The first step is to reconsider the method of finding the resultant of two parallel forces in the case where these forces are equal and opposite. (*See Art.* 10.)

Let P and P be two equal and opposite parallel forces acting at the points A, B, of a rigid body.

According to the reasoning, the resultant of P, P, is P — P, or zero, and $A D = \dfrac{Q\,a}{P-P} = \dfrac{Q\,a}{O}$.

This expression has no numerical representation, and we infer that two equal and opposite parallel forces have no resultant, in the sense that they are not equivalent to any single force tending to produce a motion of translation. But they have a resultant effect in their tendency to turn the body, and this turning effect is a measurable quantity.

Def. 1. A pair of equal and opposite parallel forces is called a *couple.*

Def. 2. The perpendicular distance between the forces is called the *arm* of the couple.

Def. 3. The product of either force into the arm is its *moment.*

Def. 4. A straight line drawn perpendicular to the plane of the couple, and proportional in length to its moment, is called the *axis* of the couple.

20. To show that the effect of a couple is completely represented by its axis.

1. The plane in which the forces act is perpendicular to the axis, and therefore the direction of the axis determines the plane of the couple.

2. Let the forces P, P, act at the ends of the arm A B, and let the axis cut the plane of the couple, either in A B, or A B produced, as at E, F.

Fig. 47.

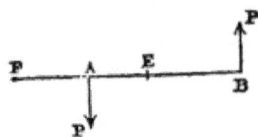

The turning effect round E
$= P \times A E + P \times E B = P \times A B.$

So also, the turning effect round F
$= P \times F B - P \times F A = P \times A B;$

and this is equally true in the extreme case when E coincides with either A or B.

Hence, the turning effect of the couple is proportional to the moment $P \times AB$, which is proportional to the axis.

Since a couple is correctly represented in all respects by its axis, it follows that we may apply to the axes of couples the same principles of composition and resolution which we have proved to be true in the case of simple forces. The so-called proposition of the parallelogram of forces may be extended to any effects completely represented by straight lines, and holds in compounding the turning effect of couples whose axes are inclined to each other.

Also we can add and subtract the parallel axes of a set of couples just as we add or subtract simple forces acting in the same straight line. By this process we obtain what is called a *resultant couple*.

21. *Prop.* To find the conditions of equilibrium of any number of forces acting in one plane on a point A.

As before, let X A X', Y A Y', be two straight lines at right angles to each other intersecting in the point A. (*Art.* 7.)

Let $P_1 P_2 \ldots$ be the forces, $a_1 a_2 \ldots$ their inclinations to X A X', and let R be their resultant, if they have one.

Then
$$X = P_1 \cos a_1 + P_2 \cos a_2 + \&c.$$
$$Y = P_1 \sin a_1 + P_2 \sin a_2 + \&c.$$
and
$$R^2 = X^2 + Y^2.$$

If there be equilibrium, $R = 0$, $\therefore X^2 + Y^2 = 0$,

$\therefore \quad X = 0, Y = 0,$

or
$$P_1 \cos a_1' + P_2 \cos a_2 + \&c. = 0 \quad \ldots \quad (1)$$
$$P_1 \sin a_1 + P_2 \sin a_2 + \&c. = 0 \quad \ldots \quad (2)$$

which are the required conditions.

22. *Prop.* To find the conditions of equilibrium of any number of forces acting in one plane upon different points of a rigid body.

Let A be any point in the body, $P_1 P_2 \ldots$ the forces acting on it, $a_1 a_2 \ldots$ the angles their directions make with a fixed line X A X'.

Let P_1 act at B, and apply at A two opposing forces, each equal and parallel to the force P_1 ; this will not disturb the equilibrium. Draw $A D_1$ perpendicular to $B P_1$ then P_1 at B

FIG 48.

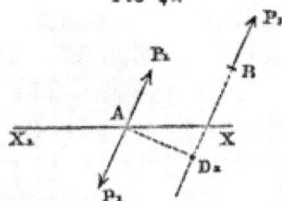

is equivalent to P_1 at A, in direction parallel to $B P_1$ and to the couple $P_1 P_1$ whose moment is $P_1 \times A D_1$.

The remaining forces P_2, P_3 . . . may be treated in like manner, and we thus obtain a collection of forces P_1, P_2, P_3 . . . acting at A in directions parallel to their actual directions, and also a collection of couples whose axes are parallel.

The couples are represented by their moments $P_1 \times A D_1$, $P_2 \times A D_2$, &c., and are equivalent to a single resultant couple whose moment is

$$P_1 \times A D_1 + P_2 \times A D_2 + \&c.$$

If there be equilibrium, the forces acting at A must be separately in equilibrium,

$$\therefore \quad P_1 \cos a_1 + P_2 \cos a_2 + \&c. = 0 \quad . \quad . \quad . \quad (1)$$
$$P_1 \sin a_1 + P_2 \sin a_2 + \&c. = 0 \quad . \quad . \quad . \quad (2)$$

Also the resultant couple must disappear ;

$$\therefore \quad P_1 \times A D_1 + P_2 \times A D_2 + \&c. = 0 \quad . \quad . \quad . \quad (3)$$

These three conditions are necessary and sufficient for the equilibrium of a rigid body under the action of forces in one plane. If any further conditions are required for the solution of a problem, they are of a geometrical character, and have nothing to do with the actions of the forces.

In the foregoing proof we have in effect stated that a force P acting at the point D, in the arm $A D$ of a lever, and tending to turn it about A, produces a push at A, which is equal to P. This is a well-known fact in mechanics, and explains the use of double lever handles, such as are commonly seen in small screw presses. If there were only a single lever handle, a pull on the lever would tend to upset the press, by reason of this transfer, whereas by putting one hand on each lever we can turn the screw without moving the frame.

THE PRINCIPLE OF ROBERVAL'S BALANCE.

23. An excellent example of the action of couples is found in the common balance where the arms are underneath the scale pans. This is Roberval's balance, and its principle will be understood from the model sketched.

Two equal bars C D, E F, are threaded on axes at the centres A and B, and are further jointed into a frame by the equal bars C E, D F.

Two bars K N, H M are attached to D F, C E, and the weights P, P are placed any-where on these bars. It is found that they balance, al-though at very unequal dis-tances from the line A B.

FIG. 49.

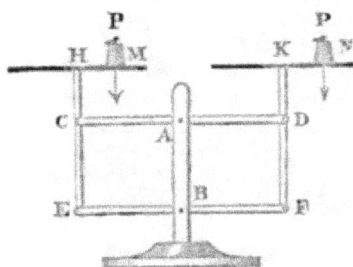

This is the very thing wanted in the balance. The scale pans are fastened to the beams, and the load is put anywhere in the pan, but the weighing is not interfered with. The balance will be perfectly accurate in its indication, although the weights are at unequal distances from the middle line. To prove this—

Apply at K two forces equal, opposite, and parallel to P, then we have in the place of P at N, a force P at D in D F, and a couple whose moment is P × N K.

Treat the force P at M in like manner and we shall replace P at M by a force P at C in C E and the couple whose moment is P × H M.

The forces P at D, and P at C balance because they are parallel and act on equal arms A D, A C.

The couples are unequal and yet they do not disturb the equilibrium : this is the peculiarity of the invention. One couple tends to twist N K, and so to pull at the peg A and to push at B, the other couple does the same, and the result is that the couples merely produce an unequal strain at A and B. They move nothing, because A and B are fixed points, and they do not interfere with the forces that are really in action.

In the actual balance the scale pans are supported at C and D, and the combination C D F E is usually concealed in the stand of the apparatus.

Ex. 1. A uniform beam A B, weighing 32 pounds and tied to E by a string E A, rests as in the figure at an angle of 30° to the horizon on a plane inclined at 60°. Find the tension of the string, and the pressure on each plane.

FIG. 50.

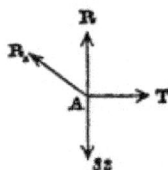
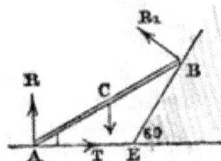

Here $T = R_1 \cos 30$, $32 = R + R_1 \sin 30$,

and $32 \times A C \cos 30 = R_1 \times A B \sin 60$, $\therefore$ $32 = 2 R_1$, $R_1 = 16$,

$$T = 16 \cos 30 = 8 \sqrt{3} = 14 \text{ nearly.}$$

$$R = 32 - \frac{R_1}{2} = 32 - 8 = 24.$$

In this example the forces 32 and R_1 are transferred to A, and two couples are introduced in consequence. The forces at A must be in equilibrium, and the couples must neutralise each other.

FIG. 51.

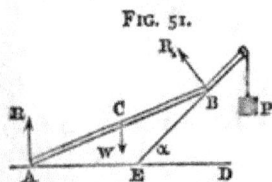

Ex. 2. The beam A B rests as in the figure, find the relation between P and W.

Let $B A D = \theta$, $B E D = \alpha$, and let $2a$ be the length of the beam, then we have

$$R_1 \sin \alpha = P \cos \alpha,$$

$$R + R_1 \cos \alpha + P \sin \alpha = W,$$

also $W a \cos \theta = R . 2 a \cos \theta$ (taking moments about B)

$\therefore$ $W = 2 R.$

$\therefore$ $\dfrac{W}{2} + \dfrac{P \cos \alpha}{\sin \alpha} \times \cos \alpha + P \sin \alpha = W,$

$$P \left(\frac{\cos^2 \alpha + \sin^2 \alpha}{\sin \alpha} \right) = \frac{W}{2}, \qquad \text{or} \qquad P = \frac{W}{2} \sin \alpha.$$

Ex. 3. A B C D is a square, a force of four units acts from A to B, a force of six units from B to C, and a force of twelve units from C to D. Find their resultant and show by a diagram how it acts.

The resultant is a force of ten units, and there is a resultant couple whose moment is 12 × BC. (Science Exam. 1872.)

Ex. 4. Let A B be a lever 8 feet long; the end A rests on a fulcrum; a weight of 40 lbs. is hung at C, three feet from A. The lever is held in a horizontal position by a force P acting vertically upwards at B. Neglect the weight of the lever and find (1) the magnitude of P, (2) the magnitude and direction of the pressure on the fulcrum.

(Science Exam. 1870.)

Here P = 15, and pressure on fulcrum = 40 − P = 40 − 15 = 25 lbs.

Ex. 5. A rod of uniform section and density weighs 10 lbs., a weight of 10 lbs. is tied to one end of it, and one of 20 lbs. to the other. Under what point of the rod must a fulcrum be placed for the whole to be in equilibrium? (Science Exam. 1871.)

Let x be the distance of the fulcrum from the end supporting 20 lbs., a the length of the rod.

$$\therefore \quad x \times 20 = 10\left(\frac{a}{2} - x\right) + 10\,(a - x) = 15\,a - 20\,x,$$

$$\therefore 40\,x = 15\,a, \qquad \therefore \frac{x}{a} = \tfrac{15}{40} = \tfrac{3}{8}.$$

Ex. 6. A uniform rod rests on two props; where must a weight, equal to the weight of the rod, be hung, so that the pressures on the props may be as two to one? *Ans.* (⅛th. from the end.)

Ex. 7. A uniform rod 6 feet long, weighing 12 lbs., and carrying a weight of 120 lbs. at a point 2½ feet from one end is supported on props at its extremities. Prove that the pressures on the props are 56 lbs. and 76 lbs.

THE USE OF A COMPENSATING LEVER.

24. In illustrating the principle of the lever, we collect the most varied examples, and the student of mechanics will find that there is something to be learnt by examining anything that is done with an object. Take the case of the compensating lever used in some locomotives for equalising the pressure on the driving wheels. It might be imagined that anyone could understand at a glance the exact use of a lever with equal arms; nevertheless, it is possible that some effort may be required before the arrangement under our notice will be thoroughly appreciated.

In a locomotive engine the weight of each part rests upon the supporting wheel, not directly, but through the medium of a powerful spring.

In the diagram, A B represents the framework, and the pressures of nine and four tons are severally communicated to the driving wheels by means of the springs E F and G H.

FIG. 52.

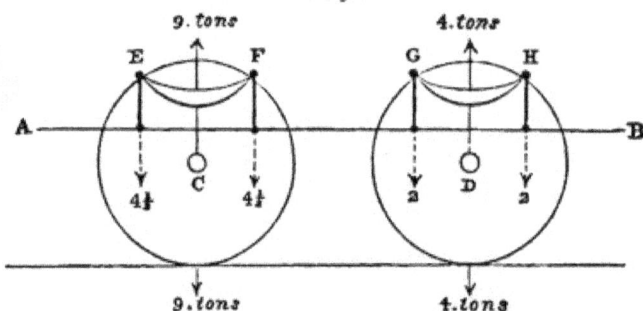

The arrows represent the directions of the forces, and since either spring, E F for example, is a lever with equal arms, acted on by two equal forces, each $4\frac{1}{2}$ tons, at its extremities, it follows that the pressure on the fulcrum will act in the vertical line through c, and be the sum of the weights pressing at E and F, that is, 9 tons.

In like manner, the forces of two tons each acting at G and H, will produce a pressure of four tons on the other wheel.

FIG. 53.

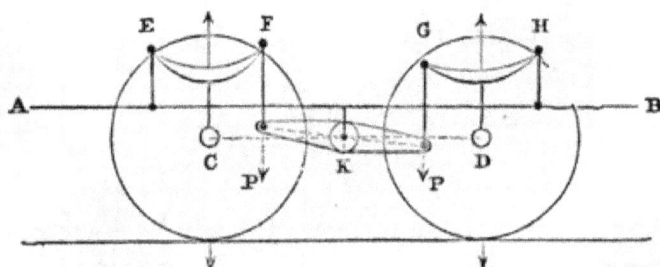

If the ends F and G, instead of being attached to the frame are connected by a lever having equal arms, and centred at K, the mechanical conditions are changed at once.

The forces at the ends of this lever must be equal, and we have two equal forces P and P, replacing the unequal forces $4\frac{1}{2}$ at F and 2 at G which were previously in action.

The force P at F, produces a pressure 2 P at C, the spring E F acting as a simple lever, whose fulcrum is E; and in like manner P at G produces a pressure 2 P at D. Hence, the weight on the rail is 2 P for each wheel, and the lever has quite abolished the previous inequality.

The student may enquire, why not assume at once that P is the mean of $(4\frac{1}{2} + 2)$ tons? As far as the drawing shows the contrivance, P would have that value; but in practice there is a third or leading wheel, and some portion of the weight may be thrown upon it by the action of the lever. It was found by trial, in the case from which the explanation is derived, that the aggregate pressure was no longer 13 tons, but was reduced to 11 tons. It is sufficient for our purpose to make out that the pressures of the two driving wheels upon the rails are made equal.

The practical value of these compensating levers is not so much to equalise the weights, since this may be done otherwise by tightening the springs, as it is to prevent jars and shocks when the engine is running upon a bad road.

Here is another illustration of the inertia of matter. It is more easy to bend the compound elongated spring formed by the two separate springs and the lever, than to divert the mass of the engine from its straight path. With a shorter and stiffer spring the shock of a sudden inequality in the road would be transferred, as by a rigid body, through the spring to the engine.

THE PRINCIPLE OF THE WHEEL AND AXLE.

25. The wheel and axle is a form of lever which allows a weight to be raised through any given height. It is a practical arrangement for continuing the action of a lever as long as may be required, the weight rising all the time.

The wheel and axle is familiar to everyone; it is used for

drawing a bucket out of a well ; the rope is wound round the axle, and the power is exerted by a lever which might be a wheel with handles, like the steering wheel of a vessel. So again, a capstan is a wheel and axle, the wheel being replaced by a set of capstan bars.

The principle of the contrivance is shown in the drawing,

Fig. 54.

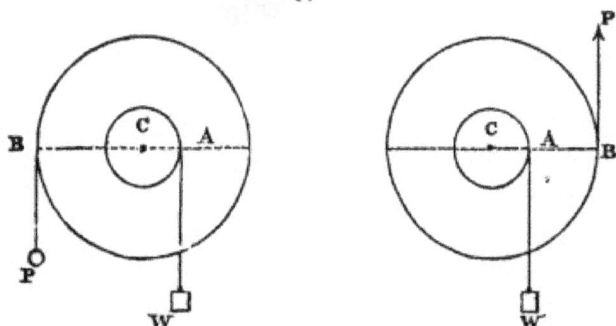

where the large circle represents the wheel, and the smaller one the axle.

The weight w hangs at A, and the power P is either a weight hanging at B, or an upward force exerted at B in a vertical direction. The latter is by far the best arrangement where it is practicable, as the pressure on C is then w—P instead of w+P, and the friction of the axle is reduced. This point will be explained hereafter.

Regarding A C B as a lever, C being the fulcrum,

we have   P × C B = W × C A,   in both cases ;

$$\text{or} \quad \frac{P}{W} = \frac{\text{radius of axle}}{\text{radius of wheel}}.$$

It is manifest that if the thickness of the rope A W be considerable, we must add half its thickness to the radius of the axle, and we must also have regard to its weight.

This is the mechanical principle of the wheel and axle, which is applied most extensively in wheelwork. As an illustration we propose to examine a method of using the wheel and axle in raising coals from deep pits.

THE WHEEL AND AXLE FOR RAISING COALS.

26. About 100 years ago the wheel and axle was employed by Smeaton for raising coals, and the apparatus was competent to raise 9 tons of coals in one hour from a depth of 600 feet.

The wheel was a double water-wheel, having two rows of buckets placed in opposite directions, and the axle was a large drum, with two ropes wound round it in opposite directions, so that one rope let down an empty tub while the other raised a full one. This is the same in principle as the modern winding engine, the only difference being in the details. A direct-acting high-pressure engine of great power, with two steam cylinders, supplies the place of the water-wheel, and acts the part of two men on a common windlass.

At the Kiveton Park Colliery, near Sheffield, the depth of the shaft is 406 yards, the wheel is a cone increasing from 20 feet to 30 feet in diameter, the cage containing the tubs of coal is drawn up in 45 seconds, and the wheel makes 14 revolutions in accomplishing this work.

It is very interesting to witness one of these gigantic wheels in action. The ease with which the steam power does its work, the perfect control over the machinery, the dial which tells the number of revolutions, the extra chalk marks on the rim of the wheel to enable the driver to pull up exactly at the right spot, the wonderful speed of the lift (three times the height of St. Paul's in as many quarters of a minute !) the splendid mechanism of the engine, the gradual starting and stopping, and the rapid swing when at full speed: we look at everything and admire the skill and courage of the practical mechanician who has brought security out of so many perils. This problem of deep winding gives a new conception of the power of the wheel and axle.

At the Rosebridge Colliery, near Wigan, the shaft is 16 feet in diameter and 815 yards deep, that is, very nearly

half a mile. The coals are raised in one minute from the
instant of starting to that of stopping. There is a point of
the descent where the steam is employed to check the
motion ; here a sudden change is felt, and the impression is
that the cage has its motion reversed and is being pulled up.
This is a curious feeling, which is mentioned because we
are about to discuss the action of the steam in bringing the
wheel safely to rest.

The problem is complicated by reason of the weight of
the rope, which here amounts to 9 lbs. per yard, and in each
revolution some 30 yards is wound up at one side and unwound at the other. This causes the load raised to vary at
every moment.

To examine the matter in its simplest form conceive that
two chains, each weighing 1 lb. per foot, are wound in oppo-

Fig. 55.

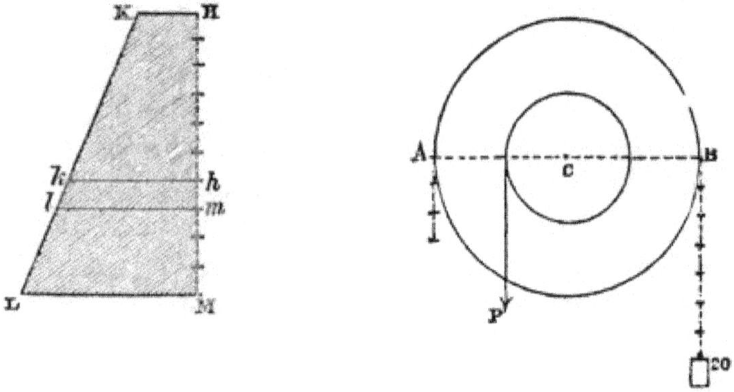

site directions on a wheel A C B, in such a manner that 10
feet of one chain overhangs when the other is wound up,
and that a weight of 20 lbs. is also to be lifted. The moment
of the load would be found as follows :—

When 3 feet is unwound at A, the load is $(20 + 7 - 3)$
or 24 lbs., and the moment of the load is 24 C B.

When 4 feet is unwound at A, the load is $(20 + 6 - 4)$
or 22 lbs., and the moment of the load is 22 C B.

When 10 feet is unwound at A, the load is $(20 - 10)$ lbs.

or 10 lbs., and the moment of the load is 10 C B. Hence
the moment of the load at each instant would be represented
by a diagram where H M is 10 feet, and the lines L M, *l m*,
k h, K H represent the forces at B when the lengths o, 3, 4,
10 feet are wound up. The sloping side K L shows the
general effect of the inequality introduced by the weight of
the rope. In order to get rid of it, the wheel is made conical, and becomes what is called a fusee. The principle of
the fusee is so important that we must examine it separately,
but it is evident that if the radius of the wheel be enlarged,
and the weight hanging becomes less, the product of the
weight into the radius of the wheel may remain constant,
and the line K L may be parallel to H M.

The drawing represents the steam cylinders employed to
drive the conical drum ; one side of the drum is filled with
the chain and the other side is empty, and since the winding
begins at the smallest part of the cone it is obviously intended
that the radius shall enlarge up to a certain point about midway in the motion.

FIG. 56.

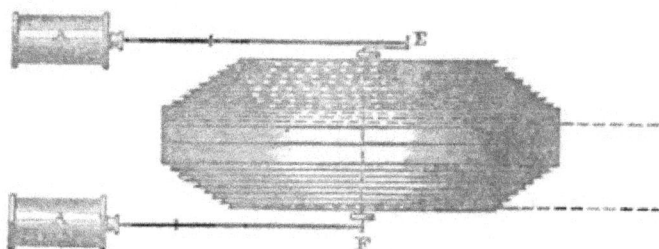

The engines act upon two cranks E and F at right angles
to each other, which replace the ordinary winch-handles ; and
the power actually exerted is that of 166 horses, the weight
of the coals raised being 22 cwts. This shows how much is
paid for the speed at which the winding is performed.

It being now made clear that the moment of the load at
any instant represents the power exerted by the load against
the rotation of the wheel, so also the moment of the steam

power on the crank levers represents the turning effect of
the engine at every instant, and may be exhibited to the
eye by a diagram. It is enormously in excess of the former
moment, and the boundary line E B C D H crosses to the other
side of the vertical during the 10th revolution. The velocity
is then so great that the cage with the coal would rise far
above the top of the shaft if the steam were not employed
with full force to bring the whole moving mass to rest. The
diagram tells us this better than any description would do,
and it is a remarkable fact that 60 tons of material in the
form of the fly-wheel, drums, ropes, cages, tubs and pulleys,
are set in motion with an estimated velocity of 36 feet per
second, in order to raise this 22 cwts. of coal. It is on
account of the great weight of the moving parts that so much
steam power is required.

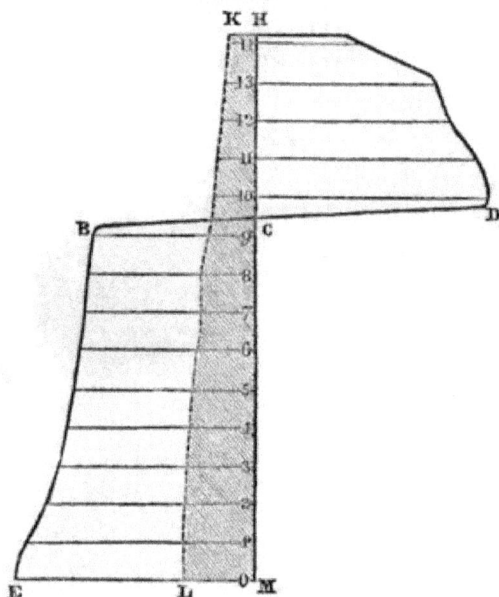

FIG. 57.

Referring to the sketch, it is seen that the shaded area informs us as to the moment of the load at any instant, and that
the effect of the conical drum is to make the line L K nearly

parallel to M H. Also E B C D H is the curve which determines the magnitude of the moment of the steam power at each instant, and the direction of the line shows that, if the first portion of this moment acts in raising the weight, the second portion will act against the weight.

The numbers 1, 2, 3 ... 14, indicate the revolutions of the drum, and the horizontal ordinates of the two curves present to the eye a comparison between the action of the load and that of the steam power at each point. The engine is reversed between the ninth and tenth revolution of the drum.

Finally it becomes a question whether the work done against the load by the reversal of the engine is really lost, or whether the momentum which so much steam power could generate is by some means stored up and preserved. We shall return to this subject when we refer to the mechanical theory of heat.

THE PRINCIPLE OF THE FUSEE.

27. The fusee is a tapering barrel used in chronometers and in most English watches, for the purpose of equalising the pull of the main-spring. The drawing shows the form of a fusee as applied in a clock train; when used in watches the fusee is shorter and there are fewer turns. The amount of taper is adjusted to the force of

Fig. 58.

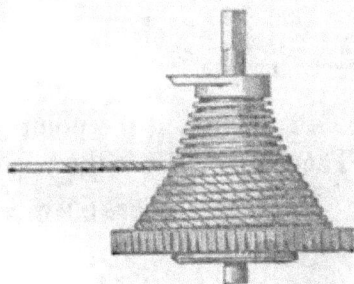

the spring by experiment, and the result is that the moment of the power to turn the fusee remains constant, although the actual pull on the chain becomes weakened as the spring uncoils. The diminishing force acts on a continually increasing arm.

28. The principle of the fusee exists wherever the arm of a lever changes continually.

Let an arm C D, moveable about C be acted on by a force

FIG. 59.

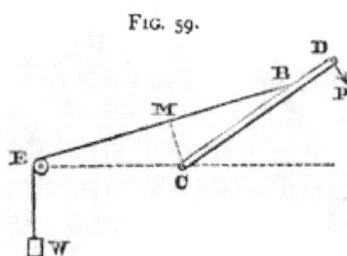

P in a direction perpendicular to the arm; and suppose a weight W to be hung on a string passing over a pulley at E and fastened to the rod at B. Draw C M perpendicular to E B. There will be equilibrium when P × C D = W × C M, or

$$\text{when} \quad W = \frac{P \times C D}{C M}.$$

If P remains constant, W will increase as C M diminishes ; or again, if W remains constant, P will become less, the more nearly C D approaches the direction of E C. In the fusee the variation of the radius of the barrel follows the variation in the force of the spring.

29. If we make E B a rigid bar, whose end E is constrained to move in the line C E, we have a combination known as the

FIG. 60.

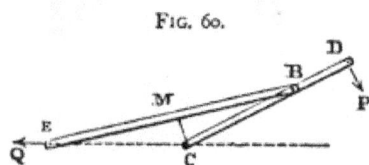

crank and connecting rod, and which also forms part of the Stanhope levers.

Let Q be the pull in direction C E exerted by a bar E B which is jointed at the point B to the bar C D.

Then Q cos C E B is the resolved part of Q in direction E B,

$$\therefore Q \cos C E B \times C M = P \times C D,$$

$$\therefore Q = \frac{P \times C D}{C M \cos C E B}.$$

In this expression nothing changes in any extreme degree except C M, which rapidly diminishes to zero as C D comes down to E C.

Hence Q increases slowly at first, but very rapidly indeed as C M approaches to zero. In fact it increases without any limit at the last instant, and the result is that this combination is extremely useful in cases where a large pressure is to be exerted through a small space, as in the printing press, or in machinery for punching metal plates.

If E B C be straightened out into a line we have another combination which is applied in an endless variety of ways.

As before, Q cos C E B is the push of Q in E B.

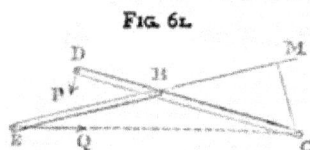

Fig. 6L.

$$\therefore \text{Q cos } C E B \times C M = P \times C D,$$

$$\therefore Q = \frac{P \times C D}{C M \cos C E B}$$

And here also Q becomes infinitely large as C M diminishes to zero, or as E B C straightens into one line.

This is often called a ' toggle joint.'

30. Among the varied instances of the applications of this combination of levers, we may point out that the toggle joint is used for supporting the head of a carriage. This joint, which is disguised in the form of an S-shaped hinged bar of iron, is grasped near the bend, and its power is shown by the ease with which the head can be straightened. When the hand is released, the head does not fall, because the angle of the joint has passed a little beyond the line joining the ends of the two arms, and the joint is constructed so as not to bend any more in that direction. The action of the head is to compress the ends and to lock the joint as soon as its angle has passed the line in which compression takes place. The tendency of the head to come down is therefore converted into the cause which prevents it from falling. The distance referred to is called the set of the joint, and would be about ⅜ inch when the joint is 24 inches long.

STONE-CRUSHING MACHINE.

31. This machine, the invention of Mr. Blake, is employed for breaking limestone and ore for blast furnaces, and is constructed so as to make use of the enormous power of the toggle joint. It is driven by steam-power, and consists of a moveable jaw C D which is hung loosely on a bar of iron at C, and vibrates a little to and fro in the direction of a fixed block. A lever A F rises and falls on the fulcrum A, and continually straightens or bends the joints E E. Each vibration of the oscillating lever may cause the lower end of the jaw to advance about $\frac{3}{8}$ inch, and then return. Upon drawing back the jaw some of the broken stone drops out, and more stone slides down ready to receive the bite of

FIG. 62.

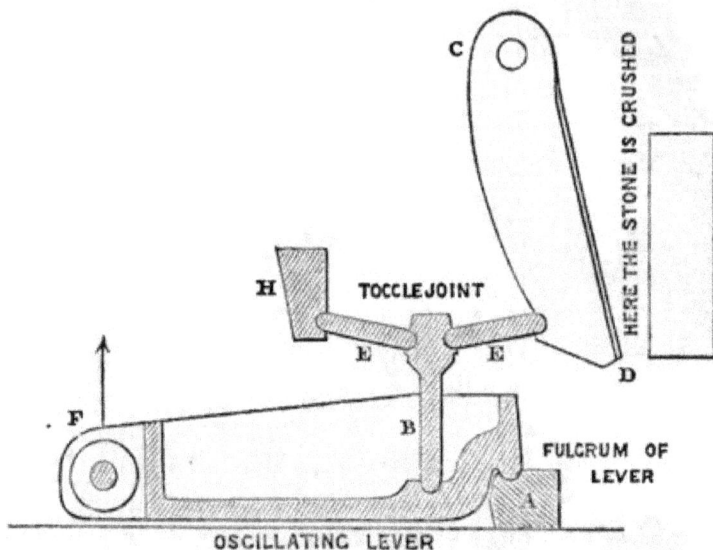

OSCILLATING LEVER

the jaw on the next oscillation. The distance between the jaws at the bottom of the opening can be regulated, and determines the size of the fragments.

This is an excellent machine and is constructed so as to be capable of breaking up 10 tons of limestone in an hour. It is largely used in granite quarries for breaking the granite chips into pieces suitable for road-making, and may be simplified by leaving out the oscillating lever.

THE STRAINS ON BEAMS.

32. The lever principle is applied in calculating the strains on girder beams and bridges.

A *transverse strain* upon a beam is produced by a force acting perpendicularly to the the direction of the beam, and tending to bend it about the point where the strain is estimated. The parts on either side of the bending point are assumed to be perfectly rigid, the bending point is taken as the centre of moments, and the strain of the beam is measured by the tendency which the forces exert to turn either portion of it round the point considered.

The transverse strain is resisted by the material of the beam, or, as in the case of a compound girder beam, by the separate action of individual lines of metal; hereafter the student will be asked to consider the method by which engineers have thrown straight line iron bridges over spans of 250 feet with perfect security.

Ex. 1. A uniform heavy beam A B, of length *l*, and weight w, is supported at its extremities; find the moment of the strain at the middle point C.

FIG. 63.

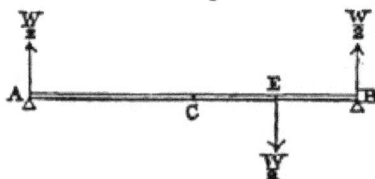

The pressure on either prop A or B is $\dfrac{w}{2}$, also the weight of C B is $\dfrac{w}{2}$, and may be supposed collected at the middle point of C B, viz., at E. Then by the principle of the lever, the moment of the strain tending to break the beam at C

$$= \frac{w}{2} \times C B - \frac{w}{2} \times C E = \frac{w}{2} \times \frac{l}{2} - \frac{w}{2} \times \frac{l}{4} = \frac{w\,l}{8}.$$

To find the moment of the strain at any other point F.

FIG. 64.

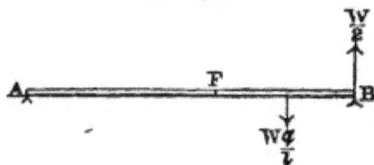

Let A F = *p*, B F = *q*, then the weight of F B $= \dfrac{w\,q}{l}$

$\therefore$ moment of strain at $F = \dfrac{W}{2} \times FB - \dfrac{W}{l}\dfrac{q}{} \times \dfrac{FB}{2}$

$$= \dfrac{Wq}{2} - \dfrac{Wq^2}{2l} = \dfrac{Wq}{2l} \times (l - q) = \dfrac{Wqp}{2l}.$$

Ex. 2. If the weight of the beam A B were supposed inconsiderable, and a weight W were hung at its middle point, we should estimate the breaking strain as follows :—

FIG. 65.

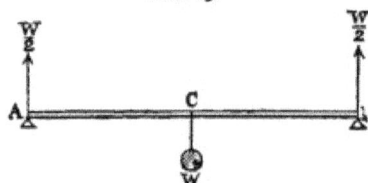

The weight w produces a pressure $\dfrac{W}{2}$ at each of the points A and B.

Hence moment of strain at $C = \dfrac{W}{2} \times CB = \dfrac{W}{2} \times \dfrac{l}{2} = \dfrac{Wl}{4}$.

Ex. 3. If the beam be loaded at any point E, and the breaking strain be required at any other point F.

FIG. 66.

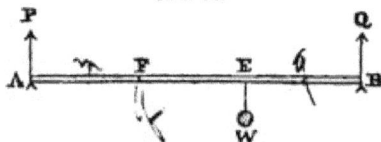

Let $AF = m$, $AE = p$, $BE = q$.

P and Q the pressures at A and B respectively.

Then   moment of strain at $F = P \times AF = \dfrac{Wq}{l} \times m = \dfrac{Wqm}{l}$.

Ex. 4. When the beam is placed obliquely, as in the common isosceles roof, the forces acting to strain it are at once deducible from the principle of the lever.

FIG. 67.

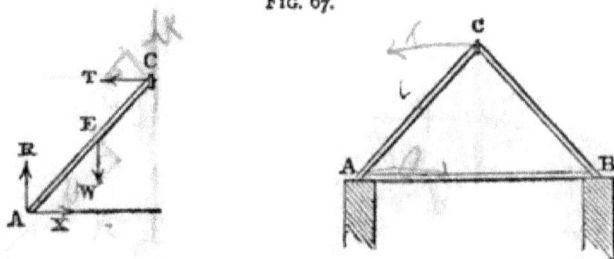

The roof consists of two equal beams A C, C B, each of weight W, and tied together by the rod A B.

The forces acting on A C are, the horizontal thrust T at C, the horizontal pull X at A, the reaction R at A, and the weight W.

Let $\qquad$ A C $= l$, C A B $= \theta$,

then $\qquad$ $T . l \sin \theta = W \dfrac{l}{2} \cos \theta$, $\quad$ taking moments about A.

$$\therefore \quad T = \frac{W}{2} \cot \theta.$$

Also, $\qquad$ R and W are the only vertical forces,

$$\therefore \quad R = W,$$

similarly $\qquad$ $X = T = \dfrac{W}{2} \cot \theta.$

Let $\qquad$ $W = 100$, $\theta = 60°$,

$$\therefore \quad \cot \theta = \frac{1}{\sqrt{3}} = \frac{1}{\sqrt{\frac{48}{16}}} = \frac{1}{\sqrt{\frac{49}{16}}} = \tfrac{4}{7} \text{ nearly,}$$

$$\therefore \quad T = 50 \times \tfrac{4}{7} = 28.8.$$

Ex. 5. A uniform beam of weight W rests on two supports, as shown in the sketch, find the tendency to break at any point.

FIG. 68.

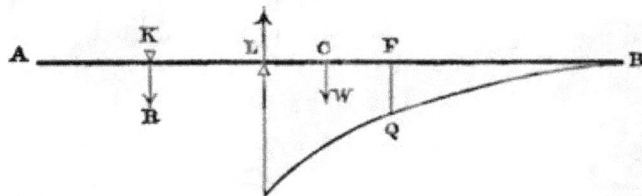

The beam is divided into three portions, in each of which the tendency changes the law of its action ; these parts are B L, L K, K A.

1. To find the tendency to break at any point F in B L.

Let $\qquad$ A B $= 2a$, B F $= x$, then weight of F B $= \dfrac{W x}{2a}$,

$$\text{moment of strain at F} = \frac{W x}{2a} \times \frac{x}{2} = \frac{W x^2}{4a}.$$

Hence the tendency to break increases as the square of the distance from B, in the manner shown by the parabolic curve B Q, where any ordinate, as F Q, represents the tendency to break at that particular point.

2. To find the tendency to break at any point E between L and K.

Let $CE = x$, $CK = c$, and let R be the pressure at K,

then $R \times KL = W \times LC$.

Also weight of $EA = \dfrac{W(a-x)}{2a}$, since $EA = a - x$.

$\therefore$ moment of strain at $E = \dfrac{W(a-x)}{2a} \times \dfrac{a-x}{2} + R(c-x)$.

$$= \dfrac{W(a-x)^2}{4a} + \dfrac{W \times LC}{KL}(c-x).$$

3. The moment of strain from A to K increases as the square of the distance from A, just as in the first case.

THE PRINCIPLE OF THE INCLINED PLANE.

33. An *inclined plane* is a plane inclined at any angle to the horizon, and the principle consists in this, that a weight W may be supported on an inclined plane by a power P which is unequal to w.   It is a direct example of the parallelogram of forces.

Prop. To find the relation between P and W on a given inclined plane.

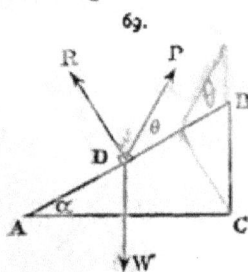

69.

Let A B be the plane; draw A C, B C horizontal and vertical lines through A and B.

Let D be a body, of weight W, supported on the plane by a force P, and let R be the reaction of the plane.

Let $BAC = a$, $PDB = \theta$.

The object is to find P and R, and there are two methods of solution which are equally convenient.

I. $\dfrac{P}{W} = \dfrac{\sin RDW}{\sin RDP} = \dfrac{\sin(90 + 90 - a)}{\sin(90 - \theta)} = \dfrac{\sin a}{\cos \theta}$,

$\dfrac{R}{W} = \dfrac{\sin PDW}{\sin RDP} = \dfrac{\sin(90 + a + \theta)}{\sin(90 - \theta)} = \dfrac{\cos(a + \theta)}{\cos \theta}$.

2. Resolving the forces along the plane and perpendicular to it, we have

$$R + P \sin \theta = W \cos a, \text{ and } P \cos \theta = W \sin a,$$

Hence $P = \dfrac{W \sin a}{\cos \theta}$.

$$\therefore R = W \cos a - \frac{W \sin a}{\cos \theta} \cdot \sin \theta,$$

$$= W \cdot \frac{\cos a \cos \theta - \sin a \sin \theta}{\cos \theta},$$

$$= \frac{W \cos (a + \theta)}{\cos \theta}.$$

Cor. 1. If P act horizontally, $\theta = - a$,

$$\therefore P = \frac{W \sin a}{\cos (-a)} = \frac{W \sin a}{\cos a} = W \tan a,$$

$$R = \frac{W \cos 0}{\cos \theta} = \frac{W}{\cos \theta}.$$

Cor. 2. If P act along the plane, $\theta = 0$,

$$\therefore P = W \sin a, \qquad R = W \cos a.$$

These corollaries may be proved independently by the triangle of forces.

1. Let P act horizontally, and draw the lines $c\,c$, $c\,b$, parallel to R, P, respectively.

Then $\dfrac{P}{W} = \dfrac{cb}{bc} = \dfrac{BC}{AC} = \dfrac{\text{height}}{\text{base}}$,

Fig. 70.

$$\frac{R}{W} = \frac{cC}{bc} = \frac{AB}{AC} = \frac{\text{length}}{\text{base}}.$$

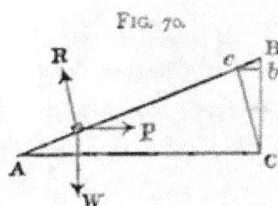

2. Let P act along the plane, and draw C E perpendicular to A B.

Then the sides of the triangle C E B are parallel to the respective forces.

$$\therefore \frac{P}{W} = \frac{BE}{BC} = \frac{BC}{AB} = \frac{\text{height}}{\text{length}},$$

$$\frac{R}{W} = \frac{EC}{BC} = \frac{AC}{AB} = \frac{\text{base}}{\text{length}}.$$

Fig. 71.

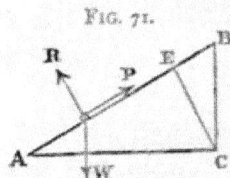

Ex. 1. If R, P, W, in the inclined plane be represented by forces of 4, 5, 7 lbs. respectively, show that $a = 44° 25'$, $\theta = 11° 32'$.

Ex. 2. A force of 40 lbs. acting parallel to an inclined plane supports 56 lbs. on the plane. The base of the plane being 340 feet, find its length and height. *Answer*, 485·8, and 346·9.

Ex. 3. A traction-engine weighs 6 tons, and is capable of drawing 20 tons over a level road ; what will it draw up a rise of 1 in 10? The coefficient of traction is 150 lbs. per ton. (Science Exam. 1872.)

34. Prop. *In an inclined plane the power* P *acts with the greatest effect when its direction is parallel to the plane.*

This is shown by taking Fig. (69) and drawing B E, C E from the points B, C respectively parallel to P and R. Then P will be least when B E is least, or when B E is perpendicular to E C, for a perpendicular is the shortest distance from the point B to the line C E.

But when B E C is a right angle, the point E falls upon A B, and the direction of P coincides with that of the plane.

CHAPTER II.

ON WORK AND FRICTION.

I⊤ has been already explained that work is done in moving a body against a resistance, and we purpose now to enter more in detail into the measure of work, and to state also the *principle of work.*

Work is done by a force when some resistance is continually overcome, and the point of application of the force is continually moved notwithstanding the resistance.

The simplest case is where the force is constant and the direction of motion is opposite to the direction of the force. The work done will then be expressed by the product of the force into the space described.

35. *The unit of work* is the work done in lifting one pound through a height of one foot, and is called a *foot-pound.*

In France the unit of work is one kilogramme raised through one metre at Paris, and is called a *kilogrammètre.* Its value is 7·2331 foot-pounds or about $7\frac{1}{4}$ foot-pounds.

The number of units of work done in a given time, say one minute, is a measure of the *efficiency* of the agent employed to do the work.

Watt estimated the work of **a horse for one minute** at 33,000 foot-pounds, and this estimate is adopted by universal consent, though it **is** too large. The number 33,000, meaning thereby 33,000 pounds raised one foot in one minute, represents a quantity of work technically termed a *Horse-Power.*

The power of a man is taken **at** 3,300 foot-pounds, or $\frac{1}{10}$th that of a horse, but it varies considerably.

The conception of work is associated with *force* and *motion*; thus force produces motion when work is being done; and accordingly it has been said that a column does no work when it supports a heavy weight, which is true; and further, that a man does no work when he supports a load on his shoulders, which appears to be quite untrue. If a man does work he gets tired, and surely it is an effort of hard **work** to support a heavy load. The truth no doubt is that the man who supports a load is continually and unconsciously lifting it through small spaces: he yields a little and recovers himself without being aware of anything more than a struggle to support the weight, and thus he does work in the true mechanical sense of the term.

Ex. How many cubic feet of water can be raised in an hour from **a** well 410 feet deep, by a steam-engine of 50 horse-power, allowing 25 per cent. for friction, leakage, &c.? (Science Exam. 1871.)

Ex. A man working on a machine performs 1,000,000 units of useful work in a day of eight hours, and the machine is so arranged that he can lift a weight of 5 cwt. How long will it take him to lift that weight through a height of 100 feet? (Science Exam. 1870.)

Ex. A man can do 900,000 units of work in a working day of nine hours, at what fraction of a horse-power does he work on the average? (Science Exam. 1872.)

Ex. It is said that a horse, walking at the rate of $2\frac{1}{2}$ miles per hour, can do 1,650,000 units of work in an hour. What force in pounds does he continuously exert? (Science Exam. 1872.)

Ex. What power of steam-engine will be required to raise 10 tons of coal per hour from a depth of 600 feet?   (Science Exam. 1871.)

Note.—The solution will give no idea of the actual power required, as no account is taken of the weight of the cages, rope, &c.

WORK DONE BY A FORCE ACTING OBLIQUELY.

36. When estimating the work done by a force which acts obliquely to the line of motion, it is sufficient to remember that a force can do no work in a line perpendicular to the direction of its action.

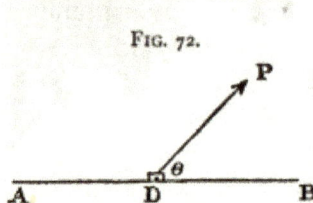

FIG. 72.

Let a force P, acting in a direction making an angle θ with A B, move a body D from A to D.

Then P cos θ, and P sin θ are the components of P, and of these, the force P cos θ does all the work.

Hence, the work done by P $= $ (P cos θ) $\times$ A D.

Prop. To find the work done by P in raising w up an inclined plane.

We refer to the diagram in Art. 33, and suppose that P acting at an angle θ to the plane A B supports a body D, of weight w.

Then the work done by P in raising D from A to B

$$= \text{P cos } \theta \times \text{A B,}$$

and the work done on D in raising it from A to B

$$= \text{w} \times \text{B C.}$$

Now it is clear that when P balances w, the work done by P is equal to the work done in moving w against the force of gravity. The force of gravity and the force P, both act upon D, and their actions in the line A P must balance each other. But if that be so, the condition of equilibrium between P and w will be that

$$\text{P cos } \theta \times \text{A B} = \text{w} \times \text{B C.}$$

Now B C $=$ A B sin a, $\therefore$ P cos $\theta \times$ A B $=$ w $\times$ A B sin a,

$$\therefore \text{ P cos } \theta = \text{w sin } a,$$

the well-known condition of equilibrium on an inclined plane.

37. This example shows us that in estimating the power of any combination it is only necessary to look at the result. The weight w is pulled up the plane by an oblique force P, yet we care not about the plane, or the force, or its direction. All we want to know is the height to which w is raised, viz., B C, and when that is given, the work done is w × B C; a certain number of foot-pounds represents this product, and the answer is complete. Furthermore, we deduce the conditions of equilibrium by simply equating the work done, under like conditions as to motion, by the separate forces.

Prop. To find the work done upon the crank in a direct acting steam-engine.

A force P, which we assume to be constant, pushes the end D of the connecting rod D B against the resistance to motion existing in the machinery connected with the crank.

FIG. 73.

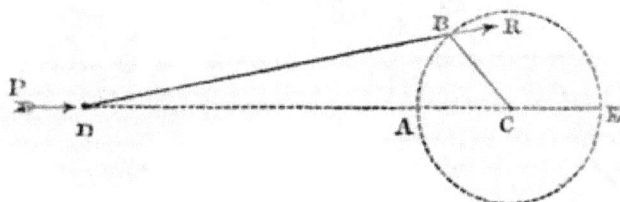

This force P produces a variable thrust R in the rod D B which turns the crank C B. What now is the work done in a half-revolution, viz., from A to E?

We are not concerned with R, either as regards its amount or direction, both are changing continually, and it would not be easy to trace their action; but we know that in the end P moves D through a space A E in the line of its action, and thereby does the work represented by P × A E.

Note. From this we conclude that there is no foundation for the popular error that some power of the steam is lost by reason of the disadvantage of the push or pull of the connecting rod when the crank C B approaches the so-called *dead points* in the line D A E.

These examples will prepare the student for a statement

of the *Principle of Work,* which is a compendious proposition containing, in its widest sense, the whole theory of equilibrium.

38. *If any system of bodies is in equilibrium under the action of forces, and we subject it to a small displacement consistent with the conditions to which the bodies are subject, the work done by all the forces is zero.*

And conversely, if the work done be zero, the forces are in equilibrium.

A large amount of ingenuity has been displayed from time to time in inventing a proof of this general proposition. Perhaps it is self-evident, but whether it be so or not, the student will now understand that it is perfectly in accord with common sense that a system of balancing forces should be regarded as incapable of doing work. Some one or more of the forces must preponderate in order that work may be done. It may suffice, therefore, to assume the truth of the principle and to give an example.

Note. *The condition of the equality of moments in a lever follows directly from the principle of work.*

Let the weights P and Q balance on a lever A C B, whose fulcrum is C, and suppose the lever to be tilted through an angle A C a, then the work done on P is $P \times a\,m$, and the work done on Q is $Q \times b\,n$.

But when P balances Q, the whole work done is zero,

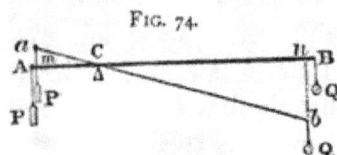

FIG. 74.

$$\therefore P \times am - Q \times bn = 0,$$
$$\text{or} \quad P \times A\,m = Q \times bn,$$
$$\text{But} \quad \frac{am}{bn} = \frac{C\,a}{C\,b} = \frac{C\,A}{C\,B},$$
$$\therefore P \times C\,A = Q \times C\,B,$$

which is the condition of equilibrium on a straight lever.

39. This principle of work includes within itself a truth which has passed into a proverb, viz., *What is gained in power* *is lost in time.* **To gain in power** means that it is possible to move a heavy **weight by a small** force, and to lose in time means that you **can only do it** very slowly.

If P be the power, and **w** the weight, x, y, the spaces moved by each respectively, we have, when P just balances w, so that no work is done in their own **directions,**

$$P\, x - W\, y = 0, \quad \text{or } P\, x = W\, y,$$

and the smallest increase in P above the value here given, will suffice to raise w ; from which it follows at once, that if w be large, and P small, x must be proportionally greater than y, or what is gained in power must be lost in the time of the movement.

THE WORK STORED UP IN A MOVING BODY.

40. We recur now to the estimation of the amount of work **stored** up in a moving body, as given in page 36. It is there shown that if a body of weight w be moving with a linear velocity v, the number of foot-pounds of work stored up in it is given by the expression $\dfrac{W\, v^2}{2\, g}$.

If a heavy particle of weight w, at a **distance** r_1 from the centre of a circle describe a circular **path with an** angular velocity ω, the amount of work stored up in it is $\dfrac{W_1\, \omega^2}{2\, g} r_1{}^2$.

The same would be true for any number of particles W_2, W_3, &c., at distances r_2, r_3, &c., the angular velocity ω being **the** same for all. But we then arrive at a solid body made **up of** parts, and rotating about an axis. Hence the work stored up in a solid body rotating about a fixed axis is given by the expression

$$\frac{\omega^2}{2\, g} \cdot (W_1\, r_1{}^2 + W_2\, r_2{}^2 + W_3\, r_3{}^2 + \text{&c.}),$$

The assistance of the mathematician is required for the summation of this series ; at present we must be content to know the fact, and to comprehend that it is possible to ascertain the exact amount of work stored up in a rotating body.

One thing may be said with advantage. The expression $\frac{1}{g}$ (w$_1$ $r_1{}^2$ + w$_2$ $r_2{}^2$ + &c.) is the sum of the products of the masses of each particle of the system into the respective squares of the distances of those particles from the axis, and we have pointed out in the introduction that the square of the distance of a particle from an axis measures its importance when revolving about that axis. Hence the above expression measures the *importance* of the matter which is rotating, or the importance of its *inertia*, and is therefore called the *moment of inertia* of the rotating body.

THE RESISTANCE OF FRICTION

41. *Friction* is the term applied to denote *the resistance to motion* which is brought into play when two rough surfaces are moved upon one another. Whatever be its origin, its power is simply that of neutralizing the action of force. It is a thing of great utility in some cases, and in others it is a source of waste and expense. Without friction an arch would not stand, a nail or a screw would be useless, and a railway train could not leave a station ; but in the parts of machinery where pieces are revolving, friction is a direct loss of power, and the object of the mechanician is to lessen it as much as possible.

42. The fundamental laws relating to friction are the following :

1. *The friction between two surfaces of the same kind is in direct proportion to the pressure between them.*

2. *The amount of friction is independent of the extent of the surfaces in contact.*

3. *The friction is independent of the velocity when the body is in motion.*

There is not space in this treatise to enter upon a discussion of the experimental proof of these laws, and we shall merely point out the methods in which they are applied.

1. Let R be the perpendicular pressure between two surfaces, F the friction, then the ratio of F to R is a constant number, when the surfaces are of the same material and in the same condition.

This number is called the *coefficient of friction*, and is denoted by the Greek letter μ, (the letter *m* of the Greek alphabet)

$$\text{that is } \frac{F}{R} = \mu, \text{ and } F = \mu R.$$

43. The *angle of repose* is the angle at which a plane A B may be inclined to the horizon before a body placed on it will begin to slide. The surfaces of the body and plane must be prepared carefully, and the forces acting are the friction F, the weight W, and the reaction R.

Draw C E perpendicular to A B, and let B A C $= a$,

FIG. 75.

Then $\dfrac{F}{R} = \dfrac{B E}{C E} = \dfrac{B C}{A C} = \tan a$,

$\therefore F = R \tan a$,

But $\quad F = \mu R$,

$\therefore \mu = \tan a$.

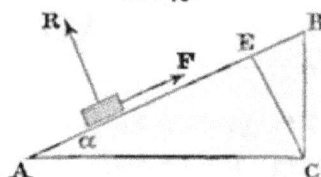

44. It is useful to record this fact by a diagram. For this purpose take a line A C $= 100$, draw C B at right angles to it, and divide C B into parts of the same size as those in A C.

If the coefficient of friction be ·08, take the 8th graduation on C E at *c* suppose, and join A *c*.

FIG. 76.

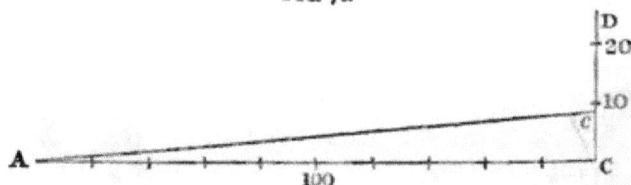

Then *c* A C represents to the eye the angle of repose.

Some of the principal results are the following:

$$\frac{C\,c}{A\,C} = \cdot 08 \text{ for metals on metals with oil.}$$

$$= .17 \text{ for metals without oil.}$$
$$= .33 \text{ for wood on wood.}$$
$$= .65 \text{ for dry masonry.}$$

Ex. 1. A body just rests without support on a plane inclined to the horizon at an angle of 30°. What is the coefficient of friction between the body and the plane ?   (Science Exam. 1871.)

Ex. 2. A body weighing 54 lbs. is just set in motion on a rough horizontal plane by a horizontal force of 9 lbs. If the force be withdrawn and the plane tilted up, at what inclination of the plane to the horizon will the body begin to slide?   (Science Exam. 1869.)

Ex. 3. A rough plane is inclined at 30° to the horizon. A weight w is placed on it, and it is found that a force $\frac{3\,W}{4}$ acting parallel to the plane will just move the weight up the plane. Find the coefficient of friction.   (Science Exam. 1870.)

Ex. 4. Two unequally rough bodies of given weights are connected by a thread and placed on an inclined plane, the less rough body being below the other. Find the inclination of the plane when the bodies begin to slide.   (Science Exam. 1873.)

THE DIRECTION OF THE REACTION OF A ROUGH SURFACE.

45. In order to form an idea of the direction in which the reaction of a rough surface acts, take the following simple experiment.

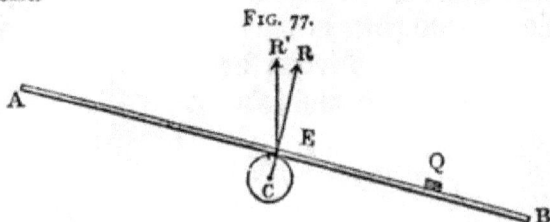

FIG. 77.

Balance a light lath on a horizontal cylinder and put a small weight Q on the lath. With care the lath may be made to tilt more and **more** till it slips off at E.

At this instant the friction is exerting its greatest effect, and it is clear that the pressure at E can never act in any other direction than the vertical line R' E drawn through E.

But action and reaction are equal and opposite, therefore the reaction at E acts in E R'.

On testing further, we should find that R E R' is *the angle of repose* for the surfaces in contact.

This experiment may be made the subject of a problem.

Let $2a$ = length of A B, b = radius of cylinder, and let Q be placed at the end B ; also let W be the weight of the lath.

At the moment when the lath is beginning to slip off we shall have $Q = \dfrac{W\, b\, a}{a - b\, a}$, where a is the angle of repose, expressed in circular measurement. It is easy to prove this by taking moments about E, and remembering that the lath rolls through an arc subtended by an angle a before it begins to slide.

46. Prop. *The direction of the reaction of a rough surface (when motion is on the point of beginning) is inclined to the perpendicular at an angle equal to the angle of repose.*

Let E be a point of some body in contact with the plane A B, and just prevented from sliding by the friction F.

Let R be the pressure at E in a direction perpendicular to A B, and R' the resultant of R and F inclined at an angle θ to E R.

Then $R' \sin \theta = F$, $R' \cos \theta = R$,

$$\therefore \quad \frac{F}{R} = \frac{R' \sin \theta}{R' \cos \theta} = \tan \theta,$$

But $\dfrac{F}{R} = \mu$. $\therefore \mu = \tan \theta$,

But $\mu = \tan a$ (*see* Art. 43.)

$\therefore \theta = a$,

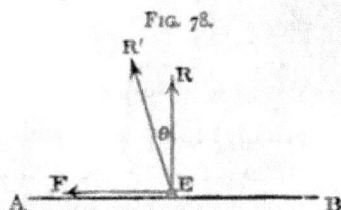

FIG. 78.

and the direction of the resultant reaction, viz. R', is inclined at an angle a to the perpendicular.

It is most necessary, in applying this proposition, to remember that it is only true when the body is on the point of sliding. The direction of the reaction of a rough surface may be strictly perpendicular to the surface, but then friction

I

is not called into play.   The effect of trying to move the
body is to incline the direction of reaction more and more
till it reaches the limit, viz. the angle of repose.

This angle, being the limit of the direction of the reaction, is frequently called the *limiting angle of resistance* of the
surface.

47. An experimental illustration is easily arranged.   Place
a slab of wood or metal on a plane surface, and press on it
obliquely by a rod.   The direction of the rod may be tilted
until it reaches the limiting angle of resistance before the
slab will move, and so long as the rod is kept within this
angle no amount of force will produce any sliding of the surfaces in contact.

The friction grips used in machinery depend on this fact.
A friction grip is obtained when any increase of pressure
causes the surfaces to bite or seize, as it is termed, more
closely.   Let E F represent a plane surface, D a slab of wood
resting on it, c another slab of wood hinged at B to a rod A B,
which is centred on an immoveable axis at A.   Now exert a

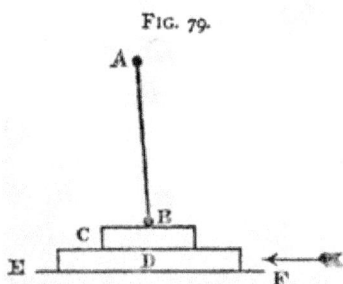

FIG. 79.

pushing force on D, it will act
through c on A B, and since A B,
on rotating round A, must press
with greater force on c, in the
line of its own direction, it
follows that so long as the
direction of A B lies within the
limits of the angle of resistance,
the attempt to move D will cause the pressure on its surfaces,
and therefore also the friction, to increase indefinitely.   But
the pressure will be powerless to cause motion for the reason
above stated, and if the friction once becomes master, it will
remain so.

The same kind of thing occurs when a drawer jams.   The
drawer must fit badly for this to happen, and should be
pulled from one side.   The drawer then twists, and two
opposite corners may bite; if they do so, the friction will

become greater, as the pull is more energetic, and very probably the drawer will not move. It can only be released by being set with the sliding surfaces quite parallel.

An example of a friction nipping lever is given in the Text Book on Mechanism.

48. Another illustration of the influence of the limiting angle is afforded by a piece of apparatus which assists in explaining the friction of an axle on its bearing. Place the axle of a tolerably heavy wheel in a bearing formed of two iron rings much larger than itself.

FIG. 80.

If the wheel be set spinning, the axle will endeavour to roll up the inside of the ring; it will rise to a certain height E, but can mount no further than the point where the surfaces begin to slip. This happens when the direction of the pressure has reached the limiting angle of resistance, or when the vertical W E makes an angle R E W, equal to the limiting angle of resistance, with the perpendicular C E R.

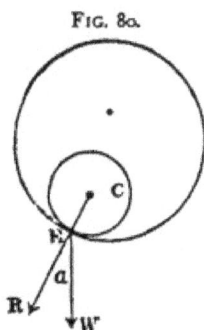

WESTON'S FRICTION COUPLING.

49. The nipping action referred to above depends on the increase of pressure, but Mr. Weston has shown that you can increase the friction as much as you please without increasing the pressure by proceeding on a different principle. In mechanics there are often several roads leading to the same result.

Conceive that a number of alternate slabs are fastened by cords to an upright piece K L, and that a number of other slabs A, B, C, are furnished with cords by which you can pull them away. Place a

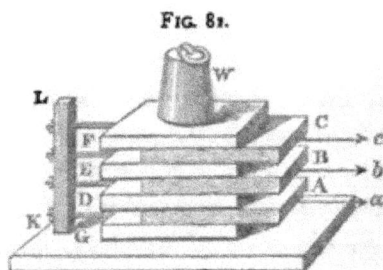

FIG. 81.

heavy weight w on the top of the pile of slabs. The friction between each pair of slabs will be the same, for we may neglect the weight of the separate slabs and regard only the weight w. If a pull is exerted on the cord A *a*, a certain resistance will be set up by friction. If two cords, as C *c* and A *a*, be pulled together, twice this resistance will be felt, and so on. But w has never varied, and therefore, without altering the pressure, it is possible to multiply the number of surfaces, and multiply the resistance of friction as much as you please.

The second law of friction, viz. that the amount of friction is independent of the extent of surfaces in contact, has been stated generally, because it is always so stated in books on mechanics. It is not true in every case, and it is not true here. But it cannot lead into error if the student will remember that it applies only where the pressure on each square inch of surface increases or decreases in the exact proportion in which the whole area of surface exposed to friction decreases or increases.

Here the pressure on each square inch is constant, whatever be the total area of surface.

There are many cases in practice where the friction is not independent of the extent of surface. Thus, the friction of a large steam piston in a cylinder is much greater than that of a small one.

50. One form of Weston's friction coupling is shown in the sketch. The object is to lock the piece c to the central shafting on which it rides ; this is done by squaring one end of the shaft at D, and threading on it a series of iron discs, whereof one is shown separately at H. In the inside of the drum are some elm-wood discs bored with a circular hole to admit the shaft D, and alternating in a series with the iron discs. The elm discs are fitted so as to rotate with the drum D, but they admit of sliding a little longitudinally in the direction of the shaft. One of these discs is shown at K.

On tightening the contact pressure between the discs of

wood and iron, by screwing up the nut furnished with a
hand wheel E, it is easy to set up a very considerable

Fig. 82.

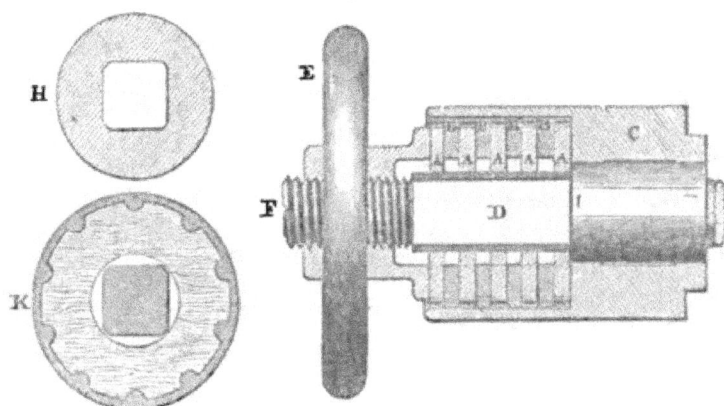

amount of friction, and thus to lock C and D together. The
apparatus is used in a 6-ton hoisting crab where C is part of
a driving pinion. When E is turned backwards, C will be
released and the weight will be lowered. When the friction
is increased the weight will run down more slowly.

The power obtained in this way is remarkable. Six discs
of iron $14\frac{1}{2}$ inches in diameter, riding between wooden discs,
and used in a windlass, are recorded to have sustained a
direct pull on the cable of 34 tons without yielding.

Ex. I. The problem of a beam resting between two inclined planes
serves very well to illustrate the observations about the reaction of a
rough surface.

Case I. Let the planes be smooth.

Then $\dfrac{AC}{CE} = \dfrac{\sin AEC}{\sin CAE}$, $\dfrac{CB}{CE} = \dfrac{\sin CEB}{\sin CBE}$.

Fig. 83.

Also $AEC = a$, $CEB = a'$,

$CAE = 90 - a - \theta$,

$CBE = 90 - a' + \theta$,

$\therefore \dfrac{\sin a}{\cos (a + \theta)} = \dfrac{\sin a'}{\cos (a' - \theta)}$,

whence $2 \tan \theta = \cot a - \cot a'$.

If the planes be rough, we have R and R', making angles ϕ and ϕ'

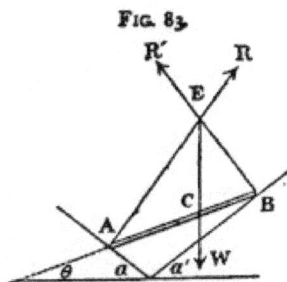

with the perpendiculars at A and B. The equations are the same as before, except that

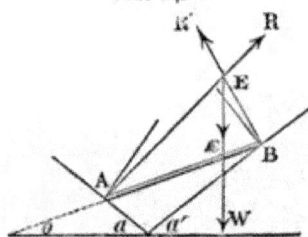
Fig. 84.

$$AEC = a + \phi, \quad CEB = a' - \phi',$$
$$CAE = 90 - a - \theta - \phi,$$
$$CBE = 90 - a' + \theta + \phi,$$
$$\therefore \frac{\sin (a + \phi)}{\cos (a + \theta + \phi)} = \frac{\sin (a' - \phi')}{\cos (a' - \theta - \phi')},$$

whence we deduce

$$\tan \theta = \cot (a + \phi) - \cot (a' - \phi').$$

Ex. 2. Find the least force which will support a weight of 100 lbs. on a plane inclined to the horizon at 30°, the force making an angle of 45° with the plane and the coefficient of friction being $\frac{1}{3}$.

Let P be the force, and a the angle of repose.

Then
$$\frac{P}{100} = \frac{\sin (30 - a)}{\sin (45 - a)} = \frac{\sqrt{2}}{2} \cdot \frac{1 - \sqrt{3} \tan a}{1 - \tan a} = \cdot 633.$$

Ex. 3. Find the least force which will drag a body of weight w along a horizontal plane.

It may be shown, by the triangle of forces, that the force is least when it makes an angle a with the plane, where a = angle of repose.

$$\therefore \quad \frac{\text{Force}}{\text{w}} = \frac{\sin (180 - a)}{\sin 90} = \sin a,$$

and force required $= \text{w} \sin a$.

THE RELATION BETWEEN HEAT AND WORK.

51. The phenomena attendant upon friction possess a peculiar interest because they have led to the development of the mechanical theory of heat. Whenever friction is overcome, heat or electricity is produced, and in dealing with masses we find that heat is the only result which is sensible. We say, therefore, that friction produces heat, and that the heat produced is an exact measure of the work expended in overcoming the friction.

According to an eminent writer, the energy of heat in the fire-box of a locomotive passes into the mechanical motion of the train, and this motion reappears as heat when the train is brought to rest. When a station is approached, the break is applied, smoke and sparks issue from it, and the momen-

tum which the train possessed is converted into the minute
and rapid vibration of heat.   •

In passing through an engineer's factory and looking at
the steel cutters at work in a lathe, we note the heat which
is set free and the jets of water which play upon the cutting
edges.   The heat evolved represents so much of the
mechanical work of the engine which is thrown away and
wasted, and this is true without any exception.   All sensible
heat produced by mechanical operations represents so much
power expended uselessly.

In the year 1849, Mr. Joule measured this loss of power,
and he concluded, from experiments made on the friction of
water, mercury, and cast-iron—

1. That the quantity of heat produced by the friction of
bodies, whether solid or liquid, is always proportional to the
quantity of work expended.

2. The quantity of heat capable of increasing the temperature of a pound of water, taken between 55° and 60°, by
1° Fahr., requires for its evolution the expenditure of a
mechanical work represented by raising 772 lbs. through a
height of one foot.

Or more concisely,

52. *Heat and mechanical work are convertible, one into the
other—viz., heat into work, or work into heat; and heat requires
for its production, or produces by its disappearance, mechanical
work in the proportion of 772 foot-pounds for each unit of heat.*

The number 772 is called the *mechanical equivalent of
heat.*   It enables us to calculate the thermal value of any
mechanical act.   Heat is a source of energy, and Mr. Joule
has given us an estimate of the amount of energy which is
stored up when a body is raised from a lower to a higher
temperature.

Def. In England the *unit of heat* is the quantity of heat
required to raise 1 lb. of water at 39·2° Fahr. by 1°.

This temperature is selected because the water has then
its maximum density.

A unit of heat is an entirely different thing from a unit of temperature. We regard heat as a source of energy. We say that the laws of motion are applicable to the case of atoms held together by molecular forces just as certainly as to bodies resting on the earth's surface. When two atoms are separated against molecular force, *potential* energy is stored up. If the atoms fall together, the potential is converted into *kinetic* energy. A quantity of heat is a quantity of energy, which may exist as potential, when it becomes latent; or as kinetic, when it becomes sensible, and influences the thermometer ; or which may exist, as is commonly the case, partly in one form and partly in the other.

Some idea of the amount of force existing in coal, by virtue of the heat which it gives out in burning, may now be gathered. The chemist, by carefully burning coal, has found that 1 lb. of ordinary coal gives out during combustion 12,000 units of heat. Now 1 unit of heat is equivalent to 772 foot-pounds of work, and therefore 12,000 units of heat represent a quantity of energy measured by 9,264,000 foot-pounds. In other words, 1 lb. of coal is capable of doing the work of 280 horses acting for 1 minute, or of 2,800 men exerting their full power during that time.

THE DUTY OF A STEAM ENGINE.

53. In the steam engine it is useful to record the work done by the burning of a given quantity of coal, and the term *duty* is applied to indicate the number of million pounds raised one foot by the burning of a bushel of coal.

In Cornwall a bushel of coals weighs 94 lbs, whereas in Newcastle it weighs 84 lbs., and the consequence is that the duty is commonly estimated with reference to the burning of 112 lbs. of coal.

Again, this measure, though suitable for estimating the work of pumping engines, is not convenient for other purposes, and the common practice now is to estimate the performance of an engine by ascertaining the number of pounds of coal burnt per hour for each horse-power at which the engine is working.

Ex. To find the duty of an engine which burn 2·2 *lbs. of coal for each horse-power per hour.*

The student should remember that a consumption at the rate of 1 lb. of coal per horse-power per hour is equivalent to a duty of 222 millions of foot-pounds.

Thus, 33,000 pounds raised 1 foot in 1 minute gives 60 × 33,000 pounds raised 1 foot in 60 minutes.

And 112 lbs. of coal will raise 112 × 60 × 33,000 pounds through 1 foot;

∴ duty of the engine = 112 × 60 × 33,000 foot-pounds.

$$= 221,760,000 \text{ foot-pounds.}$$

$$= 222 \text{ million foot-pounds nearly.}$$

If the engine burnt 2 lbs. of coal per hour its duty would be 111 million foot-pounds, and the answer required is manifestly $\frac{222}{2\cdot2}$ or about 101 million foot-pounds.

The largest amount of work yet performed by a steam-engine is a fraction under 1,000,000 pounds raised one foot by burning one pound of coal, and we have seen that one pound of coal is, *in theory*, capable of raising 9,264,000 pounds through the space of one foot.

Ex. In the ventilation of mines we have illustrations of the work done in moving large masses of air, and also of the estimation of the duty obtained from the coal.

In the Haswell Colliery, near Newcastle, the depth of the shaft is 936 feet, and it has been calculated, from observations of the temperature, that the furnace burning at the bottom of the pit raises all the air which ascends the shaft through a height of 170 feet. That is the work which it does. If the temperature of the air be 50°, and the quantity of air passing through the mine be 94,960 cubic feet per minute, what is the work done?

The weight of one cubic foot of air at 50° is ·078 lb.

∴ The weight of 94,960 cubic feet is 7,407 lbs.

This weight is lifted 170 feet in 1 minute; therefore the work done is 1,259,190 foot-pounds, which, divided by 33,000, gives 38 horse-power as the ventilating power of the furnace.

The coal consumed is 8 lbs. per minute ; hence the duty is 1,259,190 × 14 foot-pounds for 112 lbs. of coal—*i.e.* 17,628,660 foot-pounds.

THE FRICTION OF AN AXLE IN ITS BEARING.

54. We now approach a problem of the highest practical value, viz., the calculation of the work absorbed by an axle which revolves in a rough bearing.

Let any number of forces be applied to a body having a hollow cylindrical bearing surface, and resting on the fixed cylindrical axis C, and conceive that these forces are in action to press the body against the axis.

Fig. 85.

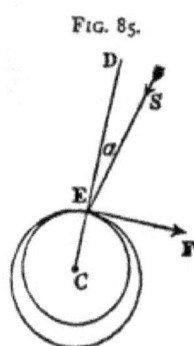

Let S be the resultant of all these forces acting at the point of contact E, F the resistance of friction, a the angle of repose.

When the body is on the point of sliding, the force S must make an angle a with the perpendicular C E D,

$$\therefore \text{F} = \text{S cos S E F} = \text{S. sin } a.$$

Let r be the radius of the axle, and let it make one complete revolution.

Then work done at E $= 2 \pi r$ S. sin a.

This is the whole theory of the subject, and the engineer has to deal with the matter so as to reduce the lost work to a minimum.

Ex. A fly-wheel weighs 20 tons, and turns on an axle 18 inches in diameter, the coefficient of friction between the axle and its bearing being 0·1. Determine approximately the number of units of work expended on friction in one turn of the wheel. (Science Exam. 1872.)

55. In discussing the problem of reducing the friction of an axle on its bearing we note that there are three quantities at our disposal, viz., S, r, and a.

First, to reduce S.

This is done by opposing as much as possible the forces which press the axle on its bearing. Take the wheel and axle as an example, and let $a =$ radius of wheel, $b =$ radius of axle, $r =$ radius of the pivot on which the axle turns.

Also let Q = weight of wheel and axle, &c., a = angle of repose, and let P be the power which is just overcoming w.

In this case s = w + Q − P, or s = w + Q + P,

according as P and w act on the same or opposite side of the centre. It is therefore a considerable advantage to arrange P and w in opposition to each other, as we see clearly from the equation of moments, which is

$$P a = w b + (w + Q − P)r \sin a.$$

If friction were abolished, we should have $P a = w b$, and we can therefore estimate the loss of power due to friction.

56. The water wheel furnishes an excellent illustration. There is a wheel at the Catrine Works in Ayrshire, which is of great power, being 50 feet in diameter, and $10\frac{1}{2}$ feet wide.

FIG. 86.

This wheel is dragged round by water descending in buckets round its periphery, and the power which it gives out is that of 120 horses. The common notion about a water wheel is that it drives machinery by means of a large toothed wheel on its axis. But let the student ask himself what would be the weight of a wheel 50 feet in diameter if it were made strong enough to resist the twisting action due to the transfer of force from the circumference to the centre. Accordingly the first substantial improvement in the construction of water wheels consisted in taking the power from the circumference of the wheel and not from its axle. When this is done the spokes of the wheel transmit no force, they merely sustain the structure; they are light tie-rods instead of being massive beams of iron.

The reduction of the weight which presses the wheel on its axle and thereby increases the friction is the first consequence of this improvement, but it is not the only one. Another result is that the power and resistance must necessarily act in opposite directions, whereby s is diminished. This point should be well understood.

Fig. 87 shows the pinion A, which gears with the large annular wheel. The line running round part of the wheel,

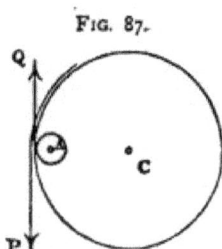

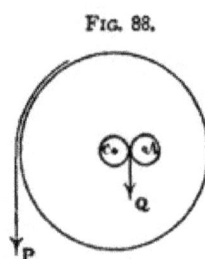

FIG. 87.   FIG. 88.

and ending in the arrow marked P, represents the pressure of the water, and the resistance to motion due to the machinery supplies the upward force Q.

If w be the weight of the wheel, we have the sum of the forces which press the wheel on its axis equal to $w + P - Q$, and the friction on the axle is $(w + P - Q) \sin a$.

Whereas in Fig. 88, where the pinion A gears with a pinion on the axle of the wheel, so placed as to disregard the reduction of friction, the reaction will occur on the

opposite side of the fulcrum c.   It follows that Q no longer
opposes w, and the friction becomes $(w + p + q) \sin a$.

This makes a very serious difference, since P and Q are
nearly equal and are both very large.   Also w requires to be
greater in the second case, and very much greater when
the wheel is of considerable dimensions.

There is yet one thing more.   The rim of a large wheel
will run at a high velocity ; in the present example any point
of it travels at a linear velocity of 4 feet per second.   The
result is that no intermediate train of wheels is required
between the water wheel and the pinion which drives the
machinery ; the high velocity renders this arrangement unnecessary, and all the friction, consequent on additional
wheelwork, is avoided.

An important step was made about 30 years ago by Sir
W. Fairbairn when he commenced to drive the shafting of
mills by teeth formed on the rim of the fly wheel of the
engine.   He did this simultaneously in both water wheels
and steam engines, and greatly simplified the motion by
imparting the requisite velocity to mill shafting from a single
wheel.   Before that time the power was taken from a spur
wheel on the axis of the engine shaft, and the requisite velocity
was obtained by adding a train of heavy wheelwork.

It follows from the principle of work that no power is lost
by driving at a greater distance from the centre with a
diminished pressure.   The loss in pressure is exactly com-
pensated by the increase in velocity.   At one time mechanical education was so imperfect that it was necessary to
argue this point.

57. Take the great beam of a steam engine ; the steam
cylinder is under one end of the beam, and the shaft of the
fly-wheel is under the other end.   The power and resistance
are on opposite sides of the fulcrum, and therefore s has its
maximum value, viz. the sum of all the forces in action.

The resistance in a pumping engine is often 40 tons, the
power is more than 40 tons, and the weight of the beam is

perhaps 35 tons ; there is, therefore, an enormous aggregate pressure producing friction. If the power and resistance acted at the same end of the beam, the pressure on the fulcrum would be little more than the weight of the beam itself. This is no doubt true, and there is a class of small engines known by the name of Grasshoppers where the power and resistance act at the same side of the fulcrum. In one type of marine engines the same thing was attempted, but it has not proved successful.

58. There is another example in the transmission of power which is familiar to every engineer. · When a pulley with a strap round it carries the motive power from one line of shafting to another, the pressure on the axis of the shaft is the sum of the tensions of the band. Probably one of these tensions, viz. the working tension, will be twice the other, on account of the absorption of force by the friction of the surface of the pulley, but still a pressure producing friction is introduced without any absolute necessity for it.

If toothed wheels were employed to convey the power, the pressure on the axis would be the resistance simply, and not the resistance added to a superfluous force. This fact does not prevent the use of bands, for they are most valuable in machinery, but may exercise some influence in practice.

59. Secondly, to diminish r.

Here it is important to distinguish between light and heavy mechanism. In watch-work, the force transmitted is so small as to be scarcely appreciable in those parts where friction comes into play—such, for instance, as the vibrating balance wheel or the escape-wheel. Accordingly the watch-maker reduces the pivots to a fine wire of steel, and only leaves material enough to support the wheel. He can do this because he has to transmit motion and not force. In delicate philosophical apparatus the use of pivots is abandoned altogether, because the friction of an axis is too serious an objection where we are concerned with minute forces. The suspension which is adopted is that of a small

fibre of silkworm's thread. This holds the light thin strip
of magnetized steel which forms the needle of a galvanometer, and substitutes an infinitesimal resistance to torsion
in the place of friction. We may note another ingeni-
ous contrivance for avoiding a mechanical difficulty. These
needles were formerly encumbered with pointers made of the
lightest material, but the inertia of which always remained
to diminish the sensitiveness of the instrument. Sir W.
Thomson fastened a very small piece of silvered glass no
thicker than paper to the magnet, and caused a beam of
light to be reflected from it. This beam may be 20 feet long,
and, when tracked in the dust of a room, is as visible as if it
were a bar of wood; but it weighs nothing, it has no inertia, and
the mirror itself is close to the axis of rotation, where the
harm it can do is the least possible.

This invention has given rise to a class of instruments of
which the reflecting mirror galvanometer is the type, and
scientific men are now enabled to observe results which
could not formerly have been made visible by any known
apparatus.

60. In light mechanism, such as the spindles and fittings of
a lathe, the system of conical bearings has superseded to a
great extent the older system of parallel bearings. The
disadvantage of conical bearings is that a circular section
near the base of the cone has a higher linear velocity than
one nearer the apex, and the wear is unequal. Accordingly
all water-taps which are made conical wear loose in time.
But setting aside this objection, it is the practice to turn
down the spindle of a small revolving piece to a conical
point and to fit this cone into a hollow coned cavity.

The radii of the rubbing surfaces are thus reduced, and the
friction is much less than it would be if the spindle were
supported on a cylindrical bearing.

The drawing shows the poppet head of a lathe; the
spindle, or mandril, which supports the speed pulley is coned
at one end in a steel point, and rests on a parallel bearing

with a shoulder at the other extremity. A small hole is

FIG. 89.

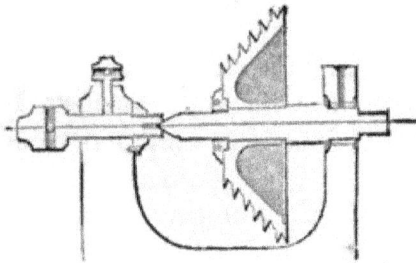

drilled truly in the axis of the mandril to receive the point of the cone, and to preserve the truth of the geometrical line as the point wears. The two systems of bearings can be examined in contrast to each other.

61. Thirdly, to diminish a.

Since a depends only on the coefficient of friction, there is little scope here for the mechanician. In watchwork it is found that the hardest steel works with the least friction against a ruby or a diamond.   The jewelling of the end of an axis of a balance or escape wheel is readily seen on examining a watch with a magnifying glass.   The pivot of steel is 'threaded through a hole bored in a ruby cup which holds the oil, and sometimes a stone cap is fitted over the pivot to prevent the endlong motion.   Again, the teeth in the scape wheel of a lever watch strike and rub against ruby plates let into the pallets.

When great force is transmitted, the bearing surface should be as soft as possible.   It will then be less easy to set up the vibratory motion of heat.   Accordingly soft metal bearings, made of tin and antimony, with a small admixture of copper, are used to form the interior lining of the brass boxes which support an axle or rubbing surface. The metal is so soft that it would be squeezed out if it were not contained in a harder casing.

Where a shaft works under water the mechanical conditions are changed, and accordingly in the case of screw-propeller shafts it has been found that hard wood bearings are superior to all others.   Here also the forces involved are very great ; the propeller itself may weigh ten or more tons, and the end of the shafting is commonly protected by a brass casing in order to get rid of any galvanic action which might injure the bearings. The shafting is therefore very ponderous and massive, and the friction has been a source of great difficulty. It is now overcome by the introduction of bearings formed of strips of lignum vitæ placed longitudinally, and forming channels for the free circulation of water. Some careful experiments have been made on the resisting power of wood, and it has been shown that wood bearings, when immersed in water, will not become abraded under a pressure of 2,000 lbs. on the square inch, and will even support an exceptional pressure of 8,000 lbs. ; whereas brass, working on iron, gives way as soon as the pressure rises to 200 lbs. on the square inch.

An indirect means of diminishing friction is to employ long bearings. A large bearing surface is essential where an axle is heavily weighted ; the pressure per square inch is not so great, and there is less danger that the film of oil between the metal surfaces will be squeezed out.

Also the bearing should be perfectly cylindrical and a very little larger than the axle which runs in it. The pressure is thus received on an adequate amount of surface, instead of on a line, as would be the case if the bearing were too large. The object is to cause the axle to float, as it were, on the thin film of oil existing between its surface and that of the cylindrical bearing. The difference of diameter will depend greatly on the fluidity of the oil employed for lubrication ; the finer the oil, the closer may be the fit. Any defect of roundness or parallelism in the axle or bearing sets up friction and heating in an inordinate degree, and thus accuracy of measurement in gauging the wearing parts, by making them true to the thousandth part of an inch, is labour well expended.

BREAK DRUMS ON THE INGLEBY INCLINE.

62. As an illustration of the effect of friction in absorbing mechanical work, and also of the importance of increasing the radius of the rubbing surface when the object is to heighten the effect of friction as much as possible, we give the following example which may be seen on a mineral branch of the North-Eastern Railway.

Here the incline is $\frac{3}{4}$ of a mile long, with an average gradient of 1 in $5\frac{3}{4}$. The wagons composing a train weigh

FIG. 90.

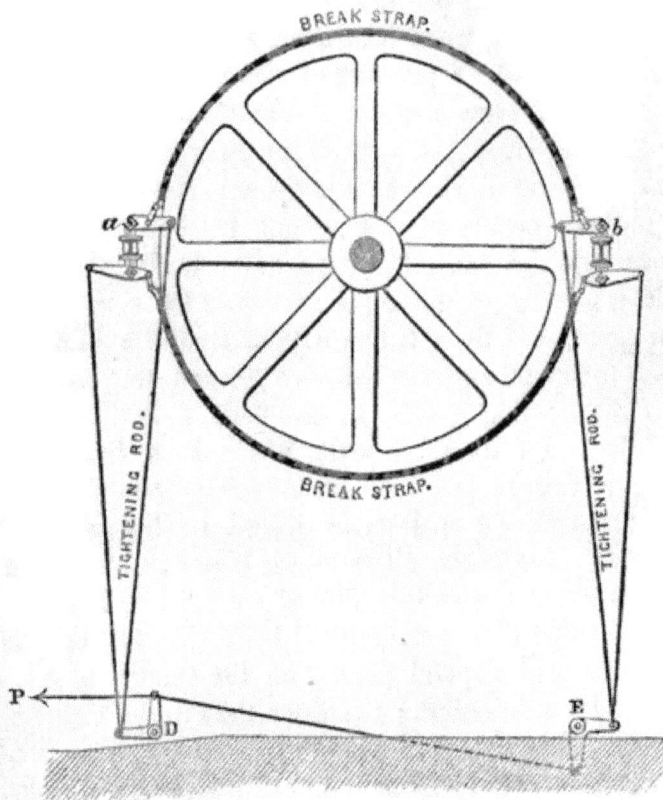

20 tons when empty, and carry 30 tons of ironstone, so that a loaded train when running down the slope is 30 tons in excess of an empty train coming up. The object of the break power is to contend against the increasing momentum of this weight of 30 tons, and to carry the train safely to the bottom of the incline. The scale of the apparatus is imposing.

On an iron shaft 15 inches in diameter are placed a pair of cast iron drums, each 18 feet in diameter, and 4 feet 8 inches broad. They are surrounded by wrought iron straps *lined with blocks of cast iron* which act like the blocks of wood in a railway break. The weight of the drums is 68 tons. The wire ropes to which the wagons are attached are 5 inches in circumference, 1,650 yards long, and weigh 8 tons.

Each run on the line occupies three minutes, the speed being 20 to 30 miles an hour, and at this rate 1,600 tons of ironstone are carried down in one day.

The mechanical point for the student is the arrangement of the break power. The strap encircling the upper half of the break wheel, is tightened by a simultaneous pull on the levers centred at *a, b*. The same thing is done with the lower half of the break strap, the bell cranks at D and E controlling both straps in the manner made clear by the drawing. Two men, acting upon a hand-winch, can work the apparatus, their power being magnified 250 times before it reaches the straps. The winch is in the direction of the arrow marked P.

The iron blocks appear to wear well, and have been substituted for wrought iron straps working on elm blocks, which speedily wore away. These elm blocks also became extremely hot, as was inevitable, the wood being a bad conductor, and the rubbing surfaces being employed in converting the energy of the moving train into heat. With iron blocks the wheel remains cool, but it must not on that account be supposed that there is any change in the action. The fact, no doubt, is that the large mass of iron takes up the heat by conduction, and dissipates it as rapidly as it is generated.

Viewing the whole operation with a knowledge of the doctrine of the conservation of energy, we observe, first, that the train at the top of the incline possesses potential energy ; secondly, that the potential is converted into kinetic energy during the descent ; and thirdly, that the surface of

K 2

the break drum takes up the kinetic energy of the moving mass, and restores it in the form of molecular motion, as a quantity of heat.

THE FRICTION OF A PIVOT.

63. Although somewhat beyond our powers of treatment in this book, it may be well to find the work absorbed by the friction of a pivot.

When an axle rests on the flat end, instead of on cylindrical bearings, it terminates in a pivot.

Let r = radius of such a pivot, w the weight supported on it, μ the coefficient of friction, and a the angle of repose. Then the area of an annulus of the pivot of radius x and breadth dx

$$= \pi (x + d x)^2 - \pi x^2 = 2 \pi x \, dx,$$
$$\text{if } (d x)^2 \text{ be neglected.}$$

Also the pressure on the annulus $= \dfrac{w}{\pi r^2} 2 \pi x \, d x$.

$\therefore$ the work absorbed by the friction on the annulus during one revolution $= \dfrac{2\,w}{r^2} . x \, dx . \mu . 2 \pi x$.

$\therefore$ the whole work absorbed by friction in one revolution of

the pivot $= \dfrac{4 \mu \pi \, w}{r^2}$. (Sum of $x^2 \, dx$).

It may be proved by analysis that the sum of a series of quantities of the form $x^2 \, dx$, where $d x$ is indefinitely small, taken between the limits $x = 0$ and $x = r$, is $\dfrac{r^3}{3}$.

Hence the work absorbed by friction in one revolution

$$= \frac{4 \mu \pi w}{r^2} . \frac{r^3}{3} = \frac{4}{3} \mu \pi w . r . = \frac{4}{3} \pi w r \tan a.$$

The loss by friction is therefore made less,

1. When w is diminished, or when the forces which press the pivot on its bearing are diminished.

2. When r is diminished.

3. When μ, or tan a, has its least value.

The method of reducing the friction of a pivot by reducing the area of the rubbing surface will be understood from the drawing, which shows the construction of the pivot of a turn-table.   In order to reduce r, and yet to preserve the pivot from rapidly grinding away, the end of the steel pin is made circular, and rests on a horizontal plate.

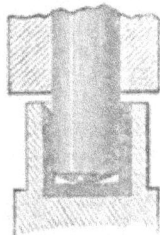

FIG. 91.
STEEL PIN

The same principle holds in philosophical apparatus.   A bar magnet may be balanced on an inverted watch-glass and supported on a glass plate.   The convex surface of the watch-glass corresponds to the rounded end of the pivot, and the magnet will rotate very readily.   A better method is to bore a hole through the centre of the magnet and cover it with a hollow agate cup.   The magnet will now balance on the point of a needle, and r is still further reduced.

THE MODULUS OF A MACHINE.

64. An endeavour has often been made to trace the work absorbed during the action of a piece of machinery, and to find out beforehand the loss traceable to friction.   A machine generally works uniformly, and when once started into full operation, the only loss is that due to friction. Mr. Moseley has proposed to express the result of the working of any machine by a formula, termed its modulus. This is a part of the analytical treatment of mechanics which is sometimes insisted upon as of the highest importance.

The object of finding the modulus is to obtain a comparison between the expenditures of moving power necessary for the production of the same effects by different machines, and the method adopted is to record the space s passed over by the point of application of the moving power in doing any work, and to obtain a relation between the work done upon the machine and the work yielded up by it.

Let u_1 be the work done by the power in moving through the space s, u_2 the work yielded up in the same time.

Then we can always obtain a relation between u_1 and u_2 in the form

$$u_1 = A\,u_2 + B\,s,$$

where A and B are constants dependent on the construction of the machine.　This relation is called its modulus.

In the wheel and axle, let P be the moving force, w the weight raised, Q the weight of the wheel, axle, and appendages.　Suppose that P is on the point of preponderating over w, and let a, b, r, be the radii of the wheel, axle, and pivot respectively, a being the angle of repose.

Then $P\,a = w\,b + (w + Q - P)\,r \sin a,$

$$\therefore P\,(a + r \sin a) = w\,(b + r \sin a) + Q\,r \sin a,$$

$$\therefore P \times 2\pi a \left(1 + \frac{r}{a}\sin a\right) = w \times 2\pi b \left(1 + \frac{r}{b}\sin a\right)$$

$$+ 2\pi a . \frac{Q\,r\sin a}{a}.$$

But $u_1 = P \times 2\pi a$, $u_2 = w \times 2\pi b$, $s = 2\pi a$, by estimating the work done in one revolution of the wheel.

$$\therefore u_1 \left(1 + \frac{r}{a}\sin a\right) = u_2 \left(1 + \frac{r}{b}\sin a\right) + s\,\frac{Q\,r\sin a}{a}$$

$$\text{or } u_1 = u_2 . \left(\frac{1 + \frac{r}{b}\sin a}{1 + \frac{r}{a}\sin a}\right) + s\,\frac{\frac{Q\,r\sin a}{a}}{1 + \frac{r}{a}\sin a}.$$

that is, $u_1 = A\,u_2 + B\,s.$

Ex. I. A single fixed pulley 2 feet in diameter turns upon an axle 1 inch in diameter, the weight of the pulley being 80 lbs.　A weight of 500 lbs. is lifted by means of this pulley ; what force is required when the coefficient of friction between the axle and its bearing is ·1 ?

(Science Exam. 1872.)

We will solve this question generally, and the student can substitute numbers in the place of symbols.

Let　w = weight raised, Q = weight of pulley,

　　　P = tension of the string, which is greater than w.

Also let a = radius of pulley, a = angle of repose,

r = radius of pin on which it rotates, here called an axle.

Then $P\,a = W\,a + (W + P + Q)\,r \sin a.$

Now when P moves through the space a, it is clear that W moves through the same space, for they hang on opposite sides of a fixed disc.

Then $u_1 = u_2 \dfrac{a + r \sin a}{a - r \sin a} + \dfrac{Q\,r \sin a}{a - r \sin a} \times s$

which gives the modulus of the *fixed* pulley.

Ex. 2. Other things being the same as before, find P when the diameter of the pulley is reduced to 6 inches, and the coefficient of friction increased to ·2. (Science Exam. 1872.)

Ex. 3. Find the modulus in the case of the single moveable pulley, the disc of the pulley having a pivot on which the friction acts.

Let a = radius of pulley, r = radius of pivot, Fig. 92.

W the weight supported, Q the weight of the pulley,

T and P the tensions of the string.

Then $T + P = W + Q,$

Also $P\,a = T\,a + (W + Q)\,r \sin a,$

$\therefore\ P\,a = (W + Q - P)\,a + (W + Q)\,r \sin a,$

or $2\,P\,a = (W + Q)\,(a + r \sin a).$

When P moves through $2\,a$, W will move through a, therefore

$$u_1 = u_2 \left(1 + \frac{r}{a} \sin a\right) + \frac{Q}{2}\left(1 + \frac{r}{a} \sin a\right) s.$$

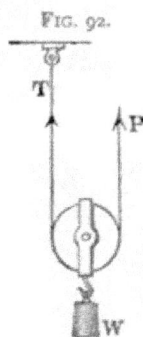

65. When a cylinder rolls on a horizontal plane, it experiences a resistance which soon brings it to rest, and which may be called rolling **resistance.** The simplest way of measuring this friction is to conceive that it opposes a *couple*, acting against the rotation of the body, and whose *moment* is the product of a very small arm, dependent on the nature of the surfaces, into the force pressing the roller against the plane. The length of this arm is given for cast iron on cast iron as ·002 of a foot (Tredgold). The amount of resistance from rolling friction is much less than that due to the sliding of two surfaces. Hence the value of friction rollers.

THE USE OF FRICTION WHEELS.

66. Friction wheels consist of **two** pairs of wheels so arranged as to form a bearing for an axle. One such pair is indicated by the circles centred at B and C, and the axle A rests in the wedge-like cavity formed by the two circumferences. Since each end of the axle is supported on a pair of wheels, a large portion of the sliding friction is eliminated, and rolling friction is substituted in its place. In light philosophical apparatus the contrivance has been very popular, and it is interesting to note the length of time during which a disc whose axle rests on friction wheels may go on rotating. But the combination is not sufficiently simple to come into general use.

Prop. To find the amount of work saved by the use of friction wheels.

Let the radii of the axles A, B, C, be r, c, c, respectively, and let a be the radius of each friction wheel. Also let C A B $= 2\,\theta$, and let S be the resultant of all the forces which press the axle A on its bearing; the direction of S being supposed to be vertical.

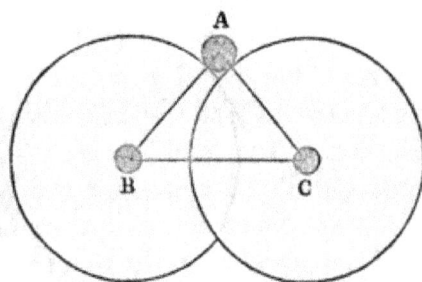

FIG. 93.

During one revolution of the axle A each friction wheel revolves through an angle ϕ, such that $\phi\, a = 2\,\pi\, r$. Also the force S produces a pressure P on the respective pivots at C and B, such that $2\,P \cos \theta = S$, and therefore the work absorbed during one revolution of the axle A is $2\,P\,c\,\phi \sin \alpha$, where α is the *angle of repose* for the surfaces in contact. Substituting for P, we have, work absorbed $= 2\,\pi\,r\,S \sin \alpha \times \dfrac{c}{a \cos \theta}$.

If there were no friction wheels, the work absorbed during one revolution of A would be $2\,\pi\,r\,S \sin \alpha$, and we conclude that the loss by friction is less than it would otherwise have been in the proportion of c to $a \cos \theta$.

ON COIL FRICTION.

67. The absorption of power when a cord is wound round a cylinder is an example of the height to which a number of extremely minute forces may rise by accumulation. The experiment is easily tried by hanging weights on a piece of cord wound round a cylindrical wooden bar, and every one must have observed the use made of it by sailors when a steamboat is being moored to a pier. The friction exerted by each portion of the rope is accumulated so as to produce the final result ; and thus, with one coil round a cylinder, the weights which balance when hung on the two ends of the rope will be in the proportion of about 1 to 9, with two coils the ratio will be about 1 to 81, and so on.

Prop. A rough cord passes over a rough cylinder and is stretched by forces at its two ends. Find the ratio between he forces when there is equilibrium.

Let P B C Q be a rough cord stretched over the surface of a rough circular cylinder whose centre is o, and leaving it at the points B and C, P and Q the forces acting on the cord, whereof Q is on the point of preponderating.

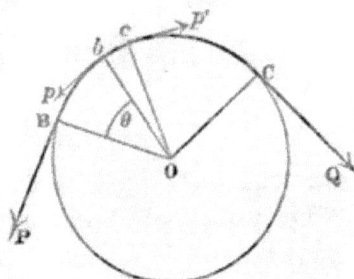

Fɪɢ. 94.

The action is this : each element of the cord adheres to the cylinder, and assists P in resisting the action of Q. It follows that the tension of the rope will vary at each point, and we shall assume it to be p, p', at the ends of a small arc bc.

Let $B O b = \theta$, $B O c = \theta'$, $B O C = \phi$, μ the coefficient of friction. It will easily be seen that the pressure on bc

$$= 2 p \cos\left(90 - \frac{\theta' - \theta}{2} \right) = 2 p \sin \frac{\theta' - \theta}{2} = p (\theta' - \theta).$$

Therefore the condition of equilibrium for the element bc is

$$p' - p = \mu p (\theta' - \theta), \text{ or } \frac{p' - p}{p} = \mu (\theta' - \theta).$$

Hence, by a well-known theorem in Algebra,*

$$\log p' - \log p = \mu \, (\theta' - \theta)).$$

That is, the increment or increase of $\log p$ is equal to μ times the increase of θ, and this being true always is true as we pass from B to C,

$$\therefore \log Q - \log P = \mu \, \phi,$$

$$\frac{Q}{P} = e^{\mu \, \phi},$$

where e is the base of the Napierian system of logarithms, and is represented by the number 2·71828. This expression shows that the difference between P and Q increases with the angle, and is independent of the radius of the cylinder.

* The theorem is the following :

$$\log \frac{z + h}{z} = \log \left(1 + \frac{z}{h} \right) = \frac{h}{z} - \frac{h^2}{2 \, z^2} + \frac{h^3}{3 \, z^3} - \&c.$$

$= \frac{h}{z}$ approximately, when $\frac{h}{z}$ is a fraction so small that its square may be neglected.   Here $\frac{p' - p}{p}$ is a similar small fraction.

THE DIAGRAM OF WORK.

68. The work done by a force can be represented to the eye by a diagram, and engineers habitually refer to such outline figures, in order to estimate at a glance the quantity of work expended in any mechanical operation.

It has been stated that the work done by a constant force is the product of the force into the space moved through by its point of application in the direction of the force.   But force and space are both represented by straight lines ; hence the work done is represented by an area.

Let A X, A Y, be two lines at right angles to each other ; call A X the line of spaces and A Y the line of resistances. Conceive that a point in a body is moved through a space M N against a constant resistance R.

Let M P = R, and complete the rectangle P N.   Then the

work done through the space M N is R × M N, or M P × M N, and is therefore represented by the rectangle P N.

If the resistance varies from N to S, according to a law indicated by the curve Q R, the work done will be represented by the area Q R S N.

Fig. 95.

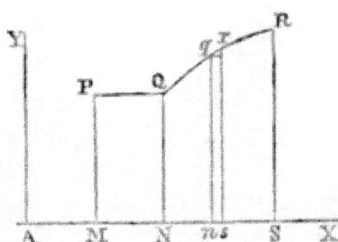

This may be proved by precisely the same reasoning as that employed in p. 30. Conceive that the space is divided into an indefinite number of equal small intervals, such as *n s*, and let *q n*, *r s* represent the resistances at *n*, *s*, respectively. If the resistance be considered constant through each small interval, the work done will be represented by the sum of a number of rectangles such as *q s*, which ultimately make up the whole area Q N S R. Hence the proposition is true.

69. *Prop.* To estimate by a diagram the work done in one revolution of the crank in a direct-acting steam-engine.

Referring to fig. 73, p. 107, let P be the thrust in D A, and let B C A = θ, B D A = φ.

Then P cos φ is the resolved part of P in direction D B, and B D makes an angle θ + φ with C B produced; therefore the components of P cos φ in directions coinciding with and perpendicular to C B are

P cos φ cos (θ + φ) and P cos φ sin (θ + φ) respectively.

In an engine, P changes at every point of the stroke, and φ also varies; but it will be useful, nevertheless, to obtain a normal diagram on the supposition that P is constant and always parallel to D A C. The components of P in and perpendicular to the crank in any position may be mapped down, and the diagram of work constructed in the manner following:—

Let A B represent an arc of the circle traced out by the crank-pin, P B the constant force r, acting on the crank C B in a direction always parallel to A C, θ the angle at which P B is inclined to the crank when in the position C B.

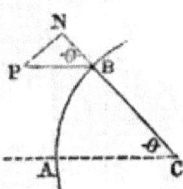
FIG. 96.

Draw P N perpendicular to C B produced; then P B may be resolved into P N and N B, that is, P $\sin\theta$ and P $\cos\theta$ respectively. Of these, P $\sin\theta$ is the component doing work, and P $\cos\theta$ produces a direct pressure on the bearings of the crank-shaft.

Next divide the circumference into a number of equal parts, and resolve the force P into its two components at each point of division; we shall then obtain a series of forces represented by the dark lines in Figs. 97 and 98.

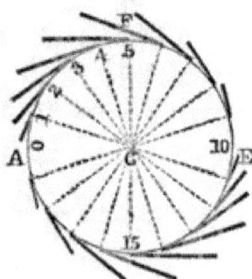
FIG. 97.

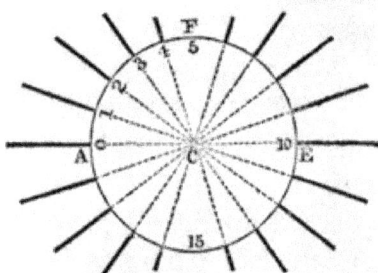
FIG. 98.

In order to construct the diagram of work, draw A F E, a line of spaces, equal in length to the semicircular arc A F E,

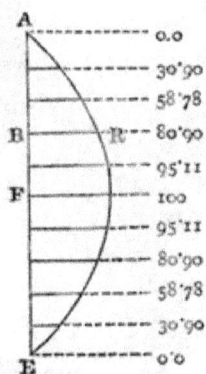
FIG. 99.

A	0.0
	30·90
	58·78
B R	80·90
	95·11
F	100
	95·11
	80·90
	58·78
	30·90
E	0·0

divide it into ten equal parts, erect perpendiculars, such as B R, equal to each corresponding component P $\sin\theta$, and complete the curve A R E. It is clear that the area A R E represents the work done on the crank in one half revolution. Let P be represented by 100, then $\sin 18° = ·3090$, and P $\sin 18 = 30·90$. That is, the values of P $\sin\theta$ for intervals of 18° will be given by the numbers in Fig. 99.

70. From what has been stated it

follows that the diagrams on pp. 92, 94, are **really** diagrams of work done.

In Fig. 55, H M is the line of spaces, whereas the perpendiculars L M, *l m*, *k h*, K H, are proportional to the pressures employed in overcoming **the** resistance of the load. The area L M H K with the sloping rectilinear side K L is **an** area representing the amount of work done in lifting the weight through the space H M. Since L K is a straight line, we infer that the resistance diminishes uniformly while the work is being accomplished.

In the same way the outline and shaded areas in Fig. 57 present to the eye an exact picture (1) of the work done by the steam, (2) of the work done directly on the load. Since the line of pressures E B D crosses to the other side of the line of spaces M C H at the point C, we infer that the area C D H represents *negative work*, or work done in opposition to that represented by the area M E B C.

Here is an example of the conversion of work into kinetic energy. The work done by the steam during about ⅔ of the motion is greatly in excess of the work done upon the load raised. Hence the steam-power is imparting a rapidly increasing velocity to the moving mass. In the closing ⅓ of the motion the steam-power is destroying the velocity already created.

If the whole of the shaded area M L K H were subtracted from the outline area M E B C, and the difference then left were compared with the outline area C D H, we should find that one is rather less than the other, and should refer the discrepancy to the fact that a certain amount of steam-power is uselessly absorbed by the friction of the moving parts. It is certain that no work can be put out of existence, and we conclude that the work done by the steam from M to C — the work done by the steam from C to H = the work done in raising the load + the work absorbed by friction.

CHAPTER III

ON THE CENTRE OF GRAVITY.

71. It may be assumed that the student is perfectly aware of the form and dimensions of the earth, and knows that the force of gravity acts in lines which may be regarded as parallel at the same place on the earth's surface. Since a body is made up of molecules or parts, and since the attraction of the earth on any one part is parallel to that on any other part, it is clear that the reasoning which led us to find the position of the centre of any number of parallel forces will apply here, and that there must exist a centre of weight which is also a centre of parallel forces.

Def. *The centre of gravity of a body is that point in which the whole weight may be supposed to act, or is the centre of parallel forces due to the weights of the respective parts of the body.*

The position of the centre may be found by experiment. It has been stated that the weight of a body is not a single force, but is, in truth, the aggregate or resultant of a series of parallel forces, and it is further clear that when a body is suspended by a string, the tension of the string must be equal and opposite to the resultant of the vertical forces due to the weights of the several parts. The string will therefore assume a vertical direction, and the centre of gravity will be found in the line of the string. Conceive that this direction is mapped down in the body itself, we shall then have a definite line passing through its centre of gravity. The point of suspension may now be changed, and a second line may be recorded, which also passes through the centre of gravity of the body. It follows that these lines will intersect, and that their point of intersection can be no other than the centre in question.

By experimenting on figures or bodies of a symmetrical form, such as a circle, a sphere, a square, or a cube, and the like, it will be found that the centre of gravity is always the centre of figure. So also the centre of gravity of a straight line is in its centre, and that of a cylindrical rod, whose ends are parallel planes, is in the centre of its axis.

The centre of gravity of many curves, areas, and solid bodies may often be determined most conveniently by analysis; but the calculations demand an advanced knowledge of mathematics, and we must therefore confine our attention to a few simple propositions.

Ex. Mention an experimental way of showing that the centre of gravity of a circular board is at its centre. (Science Exam. 1872.)

THE CENTRE OF GRAVITY OF A TRIANGLE.

72. *Prop.* To find the centre of gravity of a plane triangle.

Let A B C be a plane triangular lamina of some material, bisect A B in E, and join C E;
we shall first prove that the centre of gravity of the figure lies in C E.

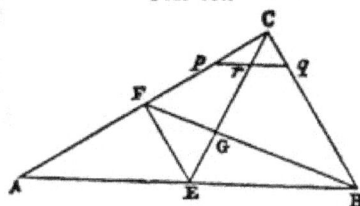

FIG. 100.

Draw any line prq parallel to A B and cutting C E in r. Then, by similar triangles prc, A E C, we have

$$pr : AE = Cr : CE.$$
In like manner $qr : EB = Cr : CE.$
$$\therefore pr : AE = qr : EB.$$
But $AE = EB \therefore pr = qr.$

or prq is bisected by C E in the point r, which is therefore its centre of gravity. The same is true of every other line drawn parallel to A B, and since the triangle is made up of an assemblage of such parallel lines or strips, each of which has its centre in C E, it follows that the centre of gravity of the triangle A B C lies in C E.

Again, bisect A C in F, and join B F, then the centre of gravity of the triangle lies in B F. But it also lies in C E, therefore it lies at their point of intersection, viz. G. Join E F. Then by similar triangles, F E G, C G B, we have

$$CG : GE = CB : FE = AB : AE = 2 : 1.$$
$$\therefore CG = 2 GE, \text{ and } CE = 3 GE,$$

whence $GE = \dfrac{1}{3} CE$, and $CG = \dfrac{2}{3} CE$.

That is, if a straight line be drawn from the angular point of a triangle to the middle of the opposite side, the centre of gravity of the triangle lies on this line at a distance from the angular point equal to ⅔ of the length of the line.

Cor. 1. The centre of gravity of three bodies of equal weight placed at A, B, C, is the same as the centre of gravity of the triangle A B C.

Let w be the weight of each of the bodies. Then the bodies w, w at A, B are equivalent to 2 w placed at E. Let G be the centre of gravity of 2 w at E, and w at C; then $2W \times EG = W \times CG, \therefore CG = 2EG$, and $CE = 3EG$, the same result as before. This establishes the corollary.

Cor. 2. If G be the centre of gravity of a triangle A B C, the forces represented in magnitude and direction by G A, G B, G C will be in equilibrium.

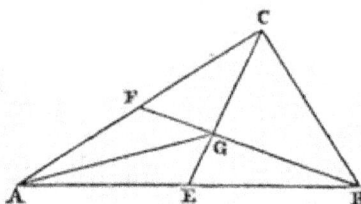

FIG. 101.

For G A is equivalent to G E, E A; and G B is equivalent to G E, E B.

$\therefore$ G A, G B are equivalent to 2 G E, E A, and E B. But E A = E B, $\therefore$ E A and E B balance each other; $\therefore$ G A, G B are equivalent to 2 G E.

But G C = 2 G E. $\therefore$ G A, G B, G C are in equilibrium when acting at the point G.

Ex. 1. A B C D is a quadrilateral figure such that the sides A B, A D, and the diagonal A C are equal, and also the sides C B and C D are equal : find its centre of gravity. (Science Exam. 1872.)

Ex. 2. A square is divided into four equal triangles by drawing its diagonals which intersect in O ; if one triangle be removed, find the centre of gravity G of the figure formed by the three remaining triangles.

Problems of this kind are frequently met with, and the method of solution is always the same.

Let O be the centre of gravity of the whole area, *g* that of the part cut out, G that of the part left. Then, by Art. 14 we are justified in concluding that (area left) × O G = (area cut out) × O *g*. In this equation everything is given except O G, which can therefore be determined.

Answer :   O G = $\frac{1}{9}$ × side of the square.

Ex. 3. **A B C** represents a triangular board weighing 10lbs. Suppose weights of 5lbs., 5lbs., and 10lbs. are placed at A, B, and C respectively. Where is the centre of gravity of the whole ?   (Science Exam. 1869.)

Ex. 4. A square board weighs 4lbs., and a weight of 2lbs. is placed at one of its corners. Show by a figure the position of the centre of gravity of the board and weight.   (Science Exam. 1871.)

Ex. 5. Find the centre of gravity of the four-sided portion of a triangle cut off by a line parallel to the base.   (Science Exam. 1868.)

THE USE OF THREE LEVELLING SCREWS.

73. Since the centre of gravity of three equal weights placed in the angular points of a triangle is the same as that of the triangle, it follows that if a heavy triangular slab be supported at its angles the pressure on each prop will be $\frac{1}{3}$ the weight of the slab.

A triangular table, standing on three legs at its angular points, is the simplest form of any, and the pressure on each leg is the same. This fact has not been without its influence in applied mechanics.

Pieces of philosophical apparatus, such as galvanometers, plates for theodolites, portable transit telescopes, and suchlike instruments, which require to be carefully levelled in a horizontal plane, are made to stand on three levelling screws. These screws lie in the points of a triangle A B C, whereby, if C be raised or lowered, the table turns about A B as an axis ; the same is true for the other points, and the levelling becomes quite easy. In some instances it is still the practice to use four levelling screws instead of three, but the manipulation is not so easy.

In the surface plate, or true plane, by Sir J. Whitworth, the same thing holds good. It is most necessary to support this plane firmly on its bearings without any strain, and therefore it stands on three angular points. The drawing

FIG. 102.

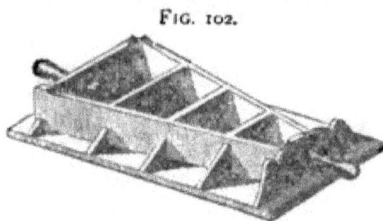

shows such a plate inverted; the back is ribbed, two supports lie near one handle, and one at the opposite side.

When a table rests on three legs and is loaded anywhere on its surface, it is easy to calculate the exact pressure on each support. When there are four legs, the p oblem is indeterminate.

Ex. A weight W is placed at any point O upon a triangular table A B C (supposed without weight). Show that the pressures on the three props, viz. A, B, C, are proportional to the areas of the triangles B O C, A O C, A O B respectively.

Draw the straight lines A O F, B O H, C O E through the point O, and let P, Q, R be the pressures at A, B, C, respectively.

Let the area $AOE = a$, $ACE = b$, $BOE = c$, $BCE = d$.

Then $R \times CE = W \times OE$.

$$\therefore \frac{R}{W} = \frac{OE}{CE} = \frac{a}{b} = \frac{c}{d}.$$

$$\therefore \frac{R}{W} = \frac{a+c}{b+d} = \frac{AOB}{ABC}.$$

Similarly $\dfrac{P}{W} = \dfrac{COB}{ABC}$, $\dfrac{Q}{W} = \dfrac{AOC}{ABC}$

$\therefore$ P : Q : R = COB : AOC : AOB,

FIG. 103.

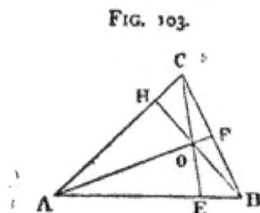

which proves the proposition, and gives a rule for ascertaining the pressure due to any given load placed on a table.

74. Another illustration is afforded by some hydraulic lifts at the Victoria Docks. Here a vessel intended to be docked is floated over a rectangular pontoon, and then lifted out of the water by a group of hydraulic presses worked by a steam-engine. There are sixteen presses on each side of the pontoon, and a ship drawing 18 feet of water can be completely raised above the surface in half an hour. The point to which we now revert is the method of supporting the great weight of the ship on a rectangular pontoon, which is, in fact, a table. The object is to avoid any unequal strain on the presses, and it is clear that if every press were worked independently, precisely the same quantity of water must be forced into each, in order to keep up a uniform lift. If all the presses were in communication, any excess of pressure on one side of the pontoon would lower that portion, and the level would not be preserved. What is wanted is a table with three supports, and that is obtained by grouping the presses into three sets. The first group is made by eight presses at the end of one side, the second group by eight presses at the corresponding end of the opposite side, and the third group by the remaining sixteen presses. Thus the three groups form a tripod stand on which the ship and pontoon rest, and the levelling is preserved throughout the operation.

THE CENTRE OF GRAVITY OF A PYRAMID.

75. *Prop.* To find the centre of gravity of a triangular pyramid.

Let A B C D represent the pyramid; bisect B C in E, and join A E, D E. Take H the centre of gravity of the triangle A B C, and join D H : we shall proceed to show that the centre of gravity of the whole pyramid lies in D H.

Let $a\,b\,c$ represent a section of the pyramid by a plane parallel to A B C, and let D H intersect $a\,b\,c$ in the point g. Draw $a\,g\,e$ meeting D E in e.

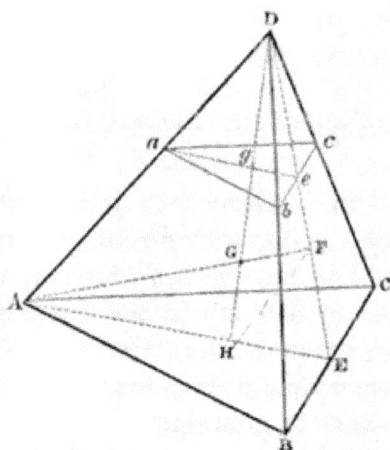

FIG. 104.

Then $a\,g : \text{A H} = \text{D}\,g : \text{D H}$,
and $\text{D}\,g : \text{D H} = g\,e : \text{H E}$,
$\therefore a\,g : \text{A H} = g\,e : \text{H E}$,
or $a\,g : g\,e = \text{A H} : \text{H E}$.
$= 2 : 1 \therefore a\,g = 2\,g\,e$.

Similarly, $b\,e : e\,c = \text{B E} : \text{E C} = 1 : 1 \therefore b\,e = e\,c$.

Hence g is the centre of gravity of the triangle $a\,b\,e$, and in like manner every section of the pyramid made by a plane parallel to A B C has its centre in D H; therefore the centre of gravity of the pyramid lies in D H. For a like reason the centre of gravity of the pyramid lies in A F, F being the centre of gravity of the triangle B D C. Let G be the point of intersection of D H and A F, and join H F.

Then $GH : GD = HF : AD = HE : AE = 1 : 3.$

$$\therefore GH = \frac{1}{3} GD, \qquad \therefore GH = \frac{1}{4} DH.$$

That is, if the vertex be joined with the centre of gravity of the base, the centre of gravity of the pyramid is in this line at ¾ the distance from the vertex.

Cor. 1. The centre of gravity of four equal bodies in the angular points of a pyramid is also that of the pyramid.

Cor. 2. Four forces acting on G, and represented in magnitude and direction by G A, G B, G C, G D, will keep the point G at rest.

Cor. 3. The centre of gravity of any pyramid, whose base is a polygon, is found by joining its vertex with the centre of gravity of the base, and taking ¾ of this distance.

Cor. 4. The centre of gravity of a cone is a point in its axis at a distance from the vertex equal to ¾ its length.

PRINCIPLE OF THE DESCENDING TENDENCY OF THE CENTRE OF GRAVITY.

76. We have seen that the weight of a body is a definite force acting on the centre of gravity and always tending to pull it downwards. So long as this centre can move by descending it will not come to rest, and since it can never ascend under the action of gravity, the position of equilibrium will be found when the centre has come to its lowest position.

The equilibrium, thus arrived at, is distinguished as being *stable*, *unstable*, or *neutral*.

1. When a body at rest is slightly disturbed, and its centre of gravity is thereby raised, we infer that the tendency of gravity will cause that centre to descend and return to its

former position. In such a case the equilibrium is stable. A hemisphere resting on a horizontal plane is an example.

Also, a body always rests in stable equilibrium when its centre of gravity lies beneath the point on which it is supported, for in that case any disturbance must raise the centre. A compass-needle is suspended on this principle.

2. When a like disturbance depresses the centre of gravity, the contrary takes place. That centre has been lowered, and gravity will prevent it from rising to its former position. The equilibrium is therefore unstable. The difficulty of balancing a long rod on the finger is an instance of unstable equilibrium. The centre of gravity is above the point of support, and although the rod would rest when truly vertical, it will always be liable to fall away from that position, unless, by a quick motion of the finger, the point of support is brought exactly beneath the centre of gravity.

3. When any slight disturbance moves the centre of gravity along a horizontal line, the equilibrium is indifferent. A cylinder rolling on a horizontal plane is an example, every position being one of equilibrium.

If the centre of gravity ascends when you deflect it from the position of rest, it is evident that the height of the centre is less than in any of the positions into which you deflect it. Hence the centre of gravity is in the *lowest position possible when the equilibrium is stable.* Whereas, if the centre of gravity descends when you deflect it from the position of rest, the height of the centre is greater than in any of the positions into which you deflect it. Hence the centre of gravity is in the *highest position possible when the equilibrium is unstable.*

77. Prop. *A body placed on a horizontal plane will stand or fall according as the vertical through its centre of gravity falls within or without the base.*

Let G be the centre of gravity of a body having a flat base and resting on a horizontal plane. The weight of the body is a vertical force w acting through G. If the body were to

fall over, it must do so by rotating about one end of a base
line A B.   In (1) this rotation will raise G, and therefore cannot be produced by gravity alone; whereas in (2) this
rotation will cause G to descend, and the tendency of gravity
is to bring G lower if possible, or to induce the motion of
falling.   Hence in the first case the body will stand, and in
the second case it will fall over.   If G be vertically above A

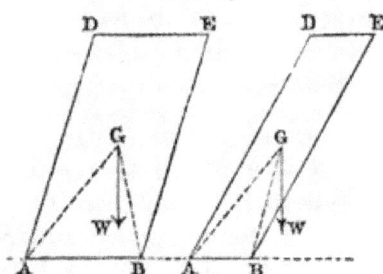

FIG. 105.

or B, the body will stand,
but the slightest disturbance will upset it.

If the plane be inclined,
the vertical through G must
still fall within the base, or
equilibrium will be impossible.

A man must stand in a
vertical position in order that the centre of gravity may fall
within the limits of his feet, which form the base on which
he stands.   In carrying a load on his back he stoops forward.   A circus rider or a skater leans in towards the centre
of the curved path in which he moves, but here another
force comes into play.   In every case of motion in a curve
the tendency to go forward in a straight line has to be overcome by a force directed towards the centre of the curve
described.   The reaction of the support acts in the line
of the man's body, which is inclined inwards, and the horizontal component of this reaction supplies the force which
enables him to preserve a curvilinear path.   For the same
reason, the outer rail, or rail farthest from the centre of
curvature, is higher than the inner one upon the curve of a
railway.

We shall proceed to determine the nature of the equilibrium when a body whose base is spherical rests on the
convex surface of a sphere.   Since the curvature of any
normal plane section of a surface at a given point is that
of its circle of curvature, the result holds for all surfaces.

78. *Prop.* A body having a spherical base is placed on the top of a sphere, to determine whether the equilibrium is stable or unstable.

The surfaces must be rough, so that one body can rock on the other, also A and B are the points originally in contact. Conceive that the body whose centre is O has rocked through a small arc B P on the sphere whose centre is C. Draw P E vertical, and let G be the centre of gravity of the upper body.

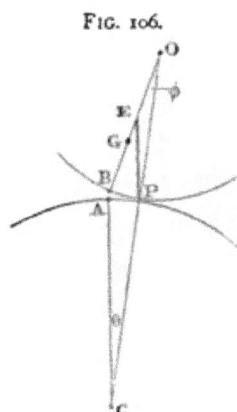

FIG. 106.

Since G always tends to get lower if possible, it is clear that the equilibrium will be stable when G falls below E.

Let $OB = r$, $AC = R$, $BOP = \phi$, $ACP = \theta$; then, by the condition of rolling, $AP = BP$, or $R\theta = r\phi$.

Also $\dfrac{OE}{r} = \dfrac{\sin EPO}{\sin OEP} = \dfrac{\sin \theta}{\sin (\theta + \phi)} = \dfrac{\theta}{\theta + \phi}$ approximately, since the angles θ and ϕ are very small.

But $\dfrac{\phi}{\theta} = \dfrac{R}{r}$, and $\therefore \dfrac{\phi + \theta}{\theta} = \dfrac{R + r}{r}$,

$\therefore \dfrac{OE}{r} = \dfrac{r}{R + r}$, or $OE = \dfrac{r^2}{R + r}$.

That is, $BE = r - OE = r - \dfrac{r^2}{R + r} = \dfrac{Rr}{R + r}$.

Let $BE = h$; then the equilibrium is stable so long as h is less than $\dfrac{Rr}{R + r}$ or $\dfrac{1}{h}$ greater than $\dfrac{1}{R} + \dfrac{1}{r}$.

Cor. 1. If the lower surface be concave, R is negative, and the equilibrium is stable when $\dfrac{1}{h}$ is greater than $\dfrac{1}{r} - \dfrac{1}{R}$.

Cor. 2. If B P be a plane, $\dfrac{1}{r} = 0$, and h is less than R.

Cor. 3. If A P be a plane, $\dfrac{1}{R} = 0$, and h is less than r.

The equilibrium is indifferent when $BE = \dfrac{Rr}{R + r}$; otherwise it is unstable, and the student can repeat the corollaries for the case of unstable equilibrium.

GENERAL METHOD OF FINDING THE CENTRE OF GRAVITY.

79. *Prop.* To find the centre of gravity of a system of particles arranged in any manner in space.

Let the system of particles A, B, C . . ., whose weights are

FIG. 107

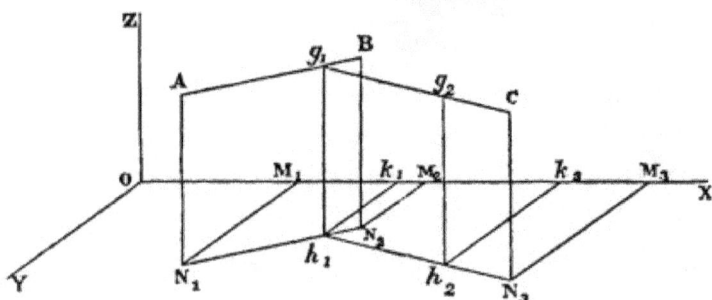

P, Q, R . . ., be referred to three lines OX, OY, OZ, mutually at right angles.

Let g_1 be the centre of gravity of A and B, g_2 that of A, B, and C, and so on. The figure is constructed by drawing A N_1 parallel to OZ (meeting the plane YOX in N_1), and $N_1 M_1$ parallel to OY, and by repeating this operation for every point g_1, g_2 . . ., and for B, C, . . . Join A B, $N_1 N_2$, g_1 C, $h_1 N_3$, . . .

Let $OM_1 = x_1$, $M_1 N_1 = y_1$, $N_1 A = z_1$, and similar quantities for B, C, . . .

Let G be the centre of gravity of the whole system, and let $\bar{x}, \bar{y}, \bar{z}$ be its co-ordinates.

Since g_1 is the centre of gravity of A and B, we have

$$P \times A g_1 = Q \times B g_1.$$

But $A g_1 : B g_1 = N_1 h_1 : h_1 N_2$

$$= O k_1 - O M_1 : O M_2 - O k_1$$

$$\therefore \mathrm{P}\,(\mathrm{o}\,k_1 - x_1) = \mathrm{Q}\,(x_2 - \mathrm{o}\,k_1).$$

$$\therefore \mathrm{o}\,k_1 = \frac{\mathrm{P}\,x_1 + \mathrm{Q}\,x_2}{\mathrm{P} + \mathrm{Q}}.$$

Similarly, $h_1 k_1 = \dfrac{\mathrm{P}\,y_1 + \mathrm{Q}\,y_2}{\mathrm{P} + \mathrm{Q}}$, and $g_1 h_1 = \dfrac{\mathrm{P}\,z_1 + \mathrm{Q}\,z_2}{\mathrm{P} + \mathrm{Q}}$.

Again, the centre of gravity of P, Q, R will be a point g_2, such that $\overline{(\mathrm{P} + \mathrm{Q} + \mathrm{R})}\,\mathrm{o}k_2 = (\mathrm{P} + \mathrm{Q})\,\mathrm{o}\,k_1 + \mathrm{R} \times \mathrm{o\,M}_3$
$$= \mathrm{P}\,x_1 + \mathrm{Q}\,x_2 + \mathrm{R}\,x_3,$$

and so on, till we arrive at the result

$$(\mathrm{P} + \mathrm{Q} + \mathrm{R} + \ldots)\,\bar{x} = \mathrm{P}\,x_1 + \mathrm{Q}\,x_2 + \mathrm{R}\,x_3 + \ldots (1)$$

Similarly,

$$(\mathrm{P} + \mathrm{Q} + \mathrm{R} + \ldots)\,\bar{y} = \mathrm{P}\,y_1 + \mathrm{Q}\,y_2 + \mathrm{R}\,y_3 + \ldots (2)$$
$$(\mathrm{P} + \mathrm{Q} + \mathrm{R} + \ldots)\,\bar{z} = \mathrm{P}\,z_1 + \mathrm{Q}\,z_2 + \mathrm{R}\,z_3 + \ldots (3)$$

The product $\mathrm{P}\,z_1$ is often called *the moment of* P *with regard to the plane* x o y. This shows an extended use of the word *moment*. The product $\mathrm{P}\,z_1$ expresses the importance of the weight of P, considered as one of a system of bodies, in affecting the position of the common centre of gravity, and the proposition proved above is equivalent to a statement that the *sum of the moments of a system of bodies with respect to any plane is equal to the moment of the whole of them* (*supposed to be collected at their centre of gravity*) *with respect to the same plane.*

Note.—The formulæ in Art. 79 are perfectly general. If $\bar{y}$ and $\bar{z}$ are each equal to zero, then (1) gives us the centre of gravity of any number of bodies arranged in a straight line. If $\bar{z} = 0$, then formulæ (1) and (2) give us the centre of gravity of any number of bodies arranged at different points in the same plane.

Ex. 1. Bodies weighing 1, 3, 5, 7 lbs. are placed at equal distances along a straight line ; find their centre of gravity.

Ex. 2. Find the centre of gravity of five equal bodies placed in the angular points of a regular hexagon.

Ex. 3. Bodies weighing 3, 8, 7, 6 lbs. are placed in this order in the angular points of a square, and a body weighing 10 lbs. is placed at the centre ; find the common centre of gravity.

THE PROPERTIES OF GULDINUS.

80. There are two remarkable properties of the centre of gravity with which we shall conclude the chapter.

I. If A B be a straight line, A P B any curve, and G the centre of gravity of this curve, regarded as a fine material line, then the area of the surface generated by the revolution of A P B about A B as an axis is found by multiplying the length of A P B into the length of the path described by G.

Let P be any point of the curve, draw P M and G H per-

FIG. 108.

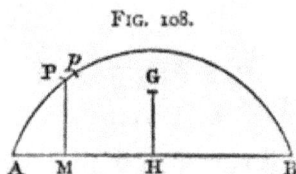

pendicular to A B, and let P p be a small arc of the curve whose length is s. Also let G H $= k$, P M $= y$, $c =$ length of path of G. Then arc described by P $: c = y : k$,

$$\therefore \text{ arc described by P} = \frac{c\,y}{k}.$$

Now the quantity of surface in the narrow strip described by P p will be found by multiplying s into the arc described by P, whence area of surface generated by $s = \dfrac{c\,y\,s}{k}.$*

Conceive that the curve is divided into an indefinite number of minute portions, which call $s, s', s'', \ldots$, at distances $y, y', y'' \ldots$ respectively from A B; then the area of the surface generated will be

$$\frac{c\,y\,s}{k} + \frac{c\,y'\,s'}{k} + \frac{c\,y''\,s''}{k} + \ldots$$

or $\dfrac{c}{k} (y\,s + y'\,s' + y''\,s'' + \ldots)$.

But $y\,s + y'\,s' + y''\,s'' + \ldots = (s + s' + s'' + \ldots)\,k$.

Hence the area of the surface generated is

$$\frac{c}{k} (s + s' + s'' + \ldots)\,k, \text{ or } c\,(s + s' + s'' + \ldots),$$

or $c \times$ length of the curve A P B.

---

* We consider P p to be so small that every point of it may be regarded as being at the same distance from A B.

2. If G be the centre of gravity of the plane area enclosed by the curved line A P B and the straight line A B, then the **volume** of the solid generated by the revolution of A P B round A B is found by multiplying the area A P B into the length of the path described by G during the revolution.

The construction and notation are the same as before, except that Q represents an indefinitely small rectangle, whose distance from A B is z, and whose area is a.

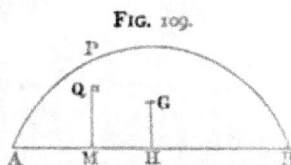

Now the whole area A P B is made up of an assemblage of very minute portions whereof the rectangle at Q is one. Also the arc described by $Q = \dfrac{cz}{k}$, and the volume of the solid generated by $Q = \dfrac{cza}{k}$. Hence the whole volume of the solid is represented by $\dfrac{c}{k}(za + z'a' + z''a'' + \ldots)$

or $\dfrac{c}{k}(a + a' + a'' + \ldots)k$ or $c \times$ area of the figure A P B.

These properties are useful in two ways. If the volume or surface of a solid be given, we can find the centre of gravity of the curve or area which generates it. Or we can apply the theorems directly to find a volume or surface when the position of the centre of gravity is known.

Ex. 1. To find the centre of gravity of the *area* of a semicircle.

Since the semicircle generates a volume $\dfrac{4}{3}\pi a^3$ by revolving about a diameter, we have $\dfrac{\pi a^2}{2} \times 2\pi \, G H = \dfrac{4\pi a^3}{3} \therefore G H = \dfrac{4a}{3\pi}$.

Ex. 2. To find the centre of gravity of the *arc* of a semicircle.

Here $\pi a \times 2\pi G H = 4\pi a^2 \therefore G H = \dfrac{2a}{\pi}$.

Ex. 3. To find the solid content of the ring of an anchor.

Let $a =$ distance from the central point of the ring to the centre of any circular section, $b =$ radius of that section

Then volume of solid $= 2\pi a \times \pi b^2 = 2\pi^2 a b^2$.

Ex. 4. To find the surface of the solid generated.

Here surface required $= 2\pi a \times 2\pi b = 4\pi^2 a b$.

CHAPTER IV.

ON SOME OF THE MECHANICAL POWERS.

81. Toothed wheels are circular discs provided with projections or teeth, which interlock as shown in the diagram,

FIG. 110.

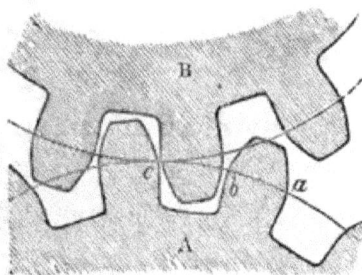

and which are therefore capable of transmitting force. When the teeth are shaped correctly, the wheels will roll upon one another, just as two ideal circles, indicated by the curved lines, and called *pitch circles*, will roll together. The pitch circle of a toothed wheel is an important element, and determines its value in transmitting motion. Suppose that two axes at a distance of 10 inches are to be connected by wheel-work and are required to revolve with velocities in the proportion of 3 to 2. Two circles, centred upon the respective axes, and having radii 4 and 6 inches, would, by rolling contact, move with the desired relative velocity, and would be the pitch circles of the wheels when made. So that whatever may be the forms of the teeth upon the wheels to be constructed, the pitch circles are determined beforehand.

The curves to be given to the teeth in order that the wheels when made shall run truly upon one another are described in the text book on Mechanism. We shall only further remark that the direction of the transmission of force between the wheels depends on the forms of the teeth, and acts in a line perpendicular to the surfaces in contact at every instant. This line is not absolutely fixed in direction, but it is immaterial for our purpose to regard its shifting character, for the teeth commonly used are so shaped that the main part of the action takes place in a line touching the two pitch circles at their point of contact.

82. The general identity of a toothed wheel with a lever may be made clear by the annexed diagram. Here a vertical bar, furnished with teeth.

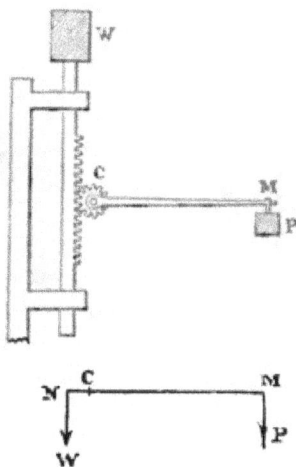

FIG. III.

supports a weight w by means of the upward pressure caused by the teeth of a wheel c, attached immoveably to a lever c m, and acted on by the power p.

The centre of the wheel is the fulcrum, one arm is c m, and the other arm is the radius of the pitch circle of the wheel. Thus the power and weight both act vertically on the lever m c n, whose fulcrum is c, and the condition of equilibrium is

$$ \text{P} \times \text{C M} = \text{W} \times \text{C N}. $$

83. The same thing is true in wheelwork generally ; each wheel of a train is merely a *mechanical equivalent* for the arm of a lever continuously in action.

Take the common case of two unequal wheels centred on the same axis, which gives the element of power in wheel-work.

Let c be the centre of motion of each of two wheels A and c in gear with other wheels D and E as shown in the diagram. The wheel D produces a pressure P acting tangentially on the circumference at A, and a pressure Q is transmitted on to E. Hence the wheels form a lever A C B with arms A C, C B, and the condition of equilibrium is

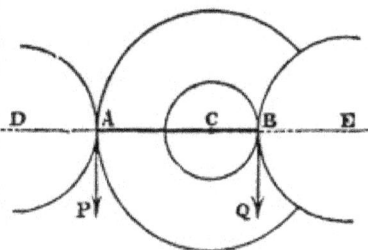

FIG. 112.

$$ \text{P} \times \text{C A} = \text{Q} \times \text{C B, or P} : \text{Q} = \text{C B} : \text{C A}. $$

That is, P : Q = circumf. of c : circumf. of A

= number of teeth in wheel c : number of teeth in wheel A.

It follows that when c = A no power is gained by the combination. In order to obtain any advantage there must be two unequal wheels on every axis.

The wheelwork of an ordinary crane furnishes an illustration, and is fairly represented in the diagram, with the exception that the wheels would be broader and more massive as we approach the drum on which the weight hangs.

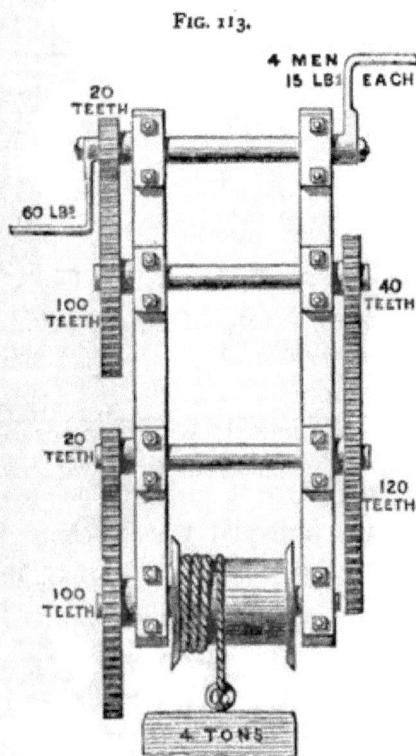

Fig. 113.

Conceive that four men, each exerting a force of 15 pounds, act upon the winch handles.

On the second axis we have the wheels 100 and 40, whereas on the third axis there are wheels 120 and 20. Each pair of wheels forms a lever, the respective arms being in the proportion of 5 : 2 and 6 : 1. Hence the force which passes through these wheels is multiplied by

$$\frac{5}{2} \times \frac{6}{1} \text{ or by } 15.$$

The winch handle and the wheel 20 form another lever, as do also the wheel 120 and the drum on which the weight hangs. Let the radius of the winch handle be $2a$ and that of the drum a, also let the radii of the wheels be $20x$ and $100x$ respectively. Then the force passing through this pair of levers is multiplied by $\frac{2a}{20x} \times \frac{100x}{a}$ or by 10.

Hence the weight raised $= 15 \times 10 \times$ power exerted,
$$= 9000 \text{ lbs.}$$

Or four men could raise somewhat more than four tons. In practice some additional force would be required in order to overcome the friction, **the** amount of which is disregarded.

THE PRINCIPLE OF REDUPLICATION.

84. This principle is applied in combinations of pulleys, where **the** pull of a stretched string comes into action several times instead **of** only once.

A *pulley-block* in a simple form **consists of two** metal plates carrying a grooved cylindrical **disc or** *sheave*. The number **of** sheaves may be increased, and **the** pulley is **then** described as a *double*, *treble*, &c. block. The drawing shows a *treble block* **in** side view **and** front elevation.

Fig. 114.

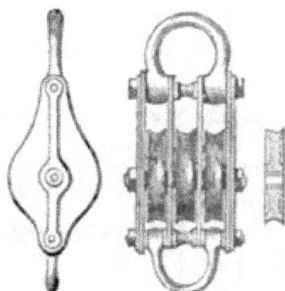

The principle of reduplication will be made clear by the statement that *a force equal to the pull of the string may be supplied at every point of departure of the string from a pulley*.

85. Take the **case of** the *single moveable pulley*, **where** P supports w, as **in the** diagram, by means **of a** string A B D P fastened at A.

Conceive **that there is no friction**, and that the string **is a fine inextensible** line destitute of weight and rigidity; the tension of the string will then be **the** same throughout, and equal to P. Since every part is at rest, we may **further** conceive that the string is deprived of **the power** of slipping, and is nailed to the pulley at the points B and D. The strings B A, D P, **will** now exert forces, each equal **to** P, **which will** be felt at the points B and D. But

Fig. 115

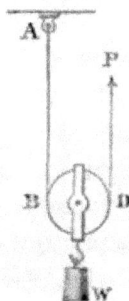

they must have exerted the same forces before the power of motion was taken away, and we conclude that the effect of coiling the string round the surface of the sheave has been simply to double the action of P in supporting w.

We have assumed that the two portions of the string are parallel and vertical, in which case $w = 2P$. If they are inclined at angle a, we have $w = 2P \cos \frac{a}{2}$.

The single moveable pulley is merely a simple lever with equal arms. The fulcrum is the centre of the pulley, and since the forces, acting on equal arms, are necessarily equal, the pressure on the fulcrum is $2P$. But w causes this pressure, therefore $w = 2P$.

86. In order to understand the action of pulleys in combination, take the forms given in the diagram. In the first case, where there are two pulleys, the pressure on A is $2P$, and that on w is $2P + P$, or $3P$. The strings attached to w are not strictly vertical, but the deviation is so small that it is not worth consideration. In the second case the tension of the string to which P is attached produces a pull on w which is manifestly $P + P + 2P$ or $w = 4P$.

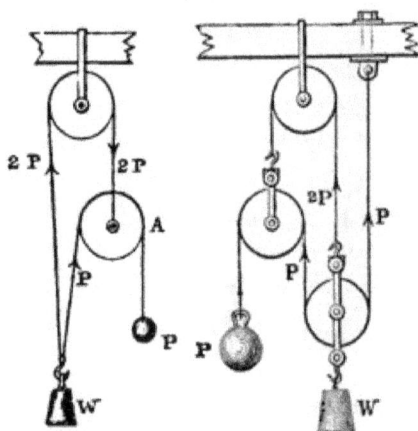

FIG. 116.

Note. The weights of the pulleys are neglected, but they can be regarded as independent weights, and it is further obvious that a comparison of the spaces through which P and w move would at once enable us to ascertain the relation between these forces.

87. As there are three kinds of levers in the books on mechanics, so there are three systems of pulleys.

1. In the first drawing, where each pulley hangs by a separate string, the pull on A is 2 P, the pull on B is 2 P + 2 P or 4 P, the pull on c is 4 P + 4 P or 8 P, that on D is 16 P, and that on E is 32 P; .

$$w = 32\,P = 2^n P,$$

where n is the number of moveable pulleys.

2. In the next drawing there are six strings on the lower block, that is, six points of departure for the string whose tension is P; therefore $w = 6P = nP$, where n is the number of strings in the lower block. The student will observe a slight deviation from the vertical in one of the strings, which is of no importance, and may be avoided by making the radii of the pulleys in the lower block proportional to 1, 3, 5, and those in the upper block proportional to 2, 4, 6.

FIG. 117.

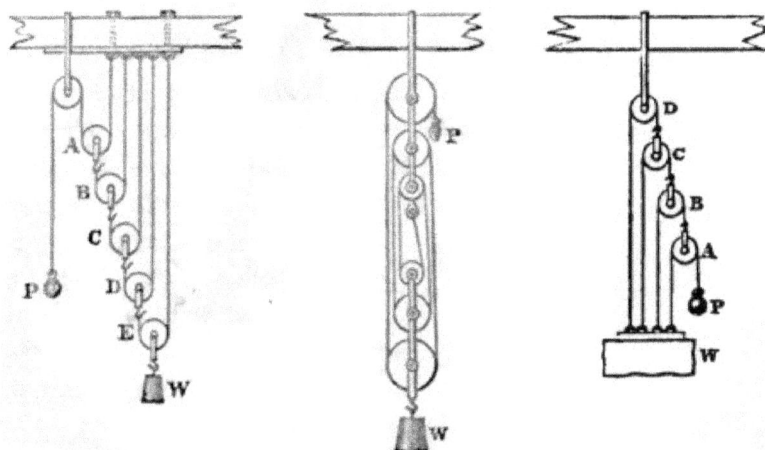

3. In the third drawing, the tension of P A is P, the tension of A B is 2P, the tension of B C is 4P, and that of C D is 8P.

$$\therefore w = P + 2P + 4P + 8P = 15P = (2^n - 1)\,P,$$

where n is the number of pulleys.

The second system is extremely valuable in practice, the only alteration being to thread the sheaves side by side on one axis, as shown in Fig. 114. The other systems are useful as exercises for the student.

Note. The pulleys are so many additional weights, assisting P in system (3), but opposing it in systems (1) and (2).

In (3), let $w_1\ w_2\ w_3\ \ldots$ be the weights of the pulleys A, B, C, beginning from the lowest pulley, and let there be n pulleys. Then

$$w = p(2^n - 1) + w_1(2^{n-1} - 1) + w_2(2^{n-2} - 1) + \ldots + w_{n-1}.$$

In (2), the weight of the pulleys is simply added to and makes part of the load w.

In (1), let $w_1\ w_2\ w_3\ \ldots$ be the weights of the pulleys E, D, C, . . . beginning from the lowest pulley, and let there be n moveable pulleys.

Then $2^n p = w + w_1 + 2w_2 + 2^2 w_3 + \ldots + 2^{n-1} w_n.$

These statements form an exercise for the student.

THE PRINCIPLE OF THE STEELYARD.

88. The steelyard is a simple lever having its fulcrum near one end. A weight fastened to a ring slides upon the longer arm, and its position indicates the weight of the substance suspended from the shorter arm. In order to exa-

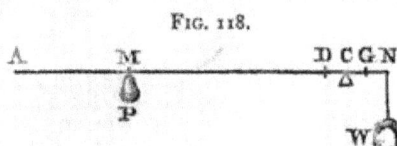
Fig. 118.

mine the principle in its simplest form, take a rigid straight line A C N to represent the steelyard, and let the moveable weight P, hung at M, balance w at N.

The first difficulty arises from the weight of the lever. The shorter arm is usually enlarged and weighted with a hook or a scale-pan, whereby C N preponderates over the arm C A, and it is necessary to hang the weight P at a point D, very near to C, in order to keep the lever horizontal. Let Q be the weight of the whole lever, together with the hook or scale-pan, G their common centre of gravity, then Q at G balances P at D, therefore $Q \times CG = P \times CD.$

Now let P at M balance w at N.

Then $P \times CM = W \times CN + Q \times CG = W \times CN + P \times CD$.

$$\therefore \quad P(CM - CD) = W \times CN; \text{ or } P = \frac{W \times CN}{DM}.$$

Let w weigh 1 pound, then $DM = \frac{CN}{P}$.

Let w weigh 2 pounds, then $DM = \frac{2CN}{P}$, and so on.

Hence the length of a division upon D A corresponding to an increase of 1 pound in the weight of w is ascertained.

In an ordinary steelyard about 20 inches long, the graduations would begin at $\frac{1}{2}$ and go up to 14 pounds. The steelyard is then turned over, and hangs from a new fulcrum much nearer to N, whereby the graduation is extended from 14 to 57 pounds. The intervals corresponding to a difference of weight equal to 1 pound are now very much smaller, C N being diminished.

89. The principle of the steelyard receives one most useful application in machinery for weighing heavy loads. The method of arrangement will be apparent from the description of the testing machine in Mr. Anderson's treatise. A drawing of the apparatus is given in page 16 of that work, and the annexed sketch shows the levers concerned in pro-

FIG. 119.

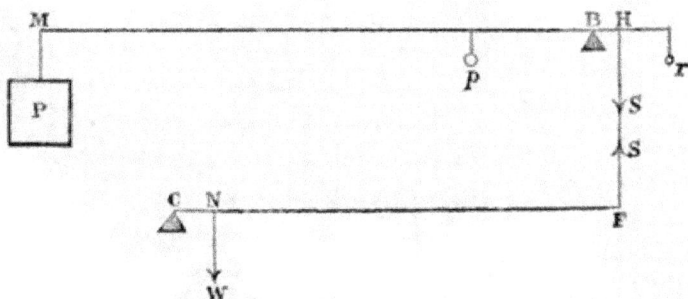

ducing the pull w. It will be seen that C F is very strong and massive, while M B is lighter and tapers towards M.

The first thing to be done is to eliminate the weights of the levers, which is effected by a moveable weight p, and a smaller weight r, both of which are so adjusted as to cause the levers to balance perfectly in a horizontal position when both P and W are removed. We may now consider the arms as rigid lines without weight.

Let $C N = 1$, $C F = 10$, then W at N is balanced by a pull S at F, such that $S \times C F = W \times C N$, or $10 S = W$.

Let $B H = 1$, $B M = 20$, then $S \times B H = P \times B M$, or $S = 20 P$.

$$\therefore \quad W = 200 P.$$

Whereby a pull of 1 cwt. at M would produce a strain of 10 tons at W, and so on.

This combination is applied for estimating the elasticity or ductility of metals. It follows that when the specimen under trial at W yields and becomes longer, the weight P will descend considerably. To get rid of this inconvenience and source of error, the fulcrum B is moveable and may be carried bodily upwards by means of a screw. This movement suffices to keep the lever H B M in a perfectly horizontal position during the operation.

THE PRINCIPLE OF THE WEDGE.

90. The wedge is a triangular prism, employed for a great variety of purposes in overcoming the force which holds the

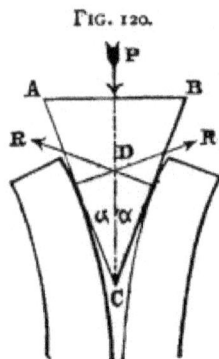

Fig. 120.

two parts of a body together. It is a double inclined plane, and is usually driven forward by a blow. The theory of its action is extremely simple, but is of no practical value, for the wedge is essentially *dynamical* in its action.

Let A B C be a smooth isosceles wedge, at rest under the action of three forces, viz., the pressure P and the reactions on the two surfaces. Since the forces balance they must meet in one point D ;

and since the wedge is isosceles, the reactions on the sides must be equal.

Let $A C B = 2a$, and let R be the reaction at either side.

Then $\dfrac{P}{R} = \dfrac{\sin R D R}{\sin R D C} = \dfrac{\sin (180 - 2a)}{\sin (90 - a)} = \dfrac{\sin 2a}{\cos a} = 2 \sin a.$

91. The application of the wedge to cutting tools is a matter of considerable interest in mechanics.

Let the student examine a tool for shaping iron; he will at once see that it is not a scraping instrument like (A), but a wedge (B), whereby each atom along the whole line of section is torn from the opposite one to which it was attached. When the shaving has been formed it runs up the slope of the wedge and is curved, bent, or broken out of the way. Further, the cutter requires to be ground as it becomes blunt, and in order to provide for this loss of material a common type of cutting tool is that shown at (C).

FIG. 121.

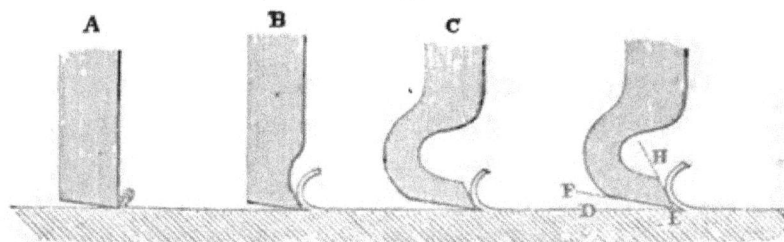

The angle D E F, between the lower edge of the material and the cutter is called *the angle of relief*, while the angle F E H is the *angle of the tool*.

It is evident that in cutting metals the angle of relief should be as small as possible, since the wedge then acts more directly. In cutting wood the angle of relief may be very much greater.

The angle F E H is ascertained by experience and is usually taken to be 60° for wrought iron, 70° for cast iron, 80° for brass; whereas for wood, a yielding material, it is much less, and ranges between 30° and 40°, being greater for the harder woods. The rate at which the wedge advances is about 10 to 15 feet per minute for cast iron, and 15 to 20 feet for wrought iron. If this speed be exceeded the cutter becomes unduly heated and the steel is injured. By a better system of lubrication the speed may be greatly increased, but it would scarcely rise under any circumstances above 100 feet per minute. Wood presents a remarkable contrast to iron in this respect, as the cutting edges used for shaping wood may travel at speeds increasing up to 7,000 or 8,000 feet per minute, and in exceptional cases, a velocity of 12,000 feet may become practicable.

But, whatever be the material under operation, or whatever may be the speed, the action of the wedge is that relied upon, and the student will be interested in observing its various modifications. Indeed, a simple cutting tool, such as a carpenter's plane, is an instrument worthy of careful examination. There are four distinct things to be done :

1. The tool removes the shaving by a *wedging* action.

2. The edge of the mouth holds down, and lessens the liability to splitting.

3. The shaving runs up the iron and is bent out of the way. When a second or breaker iron overlies the cutting iron, this bending up of the shaving is effected more perfectly and the cut is cleaner.

4. The sole, or wooden base, prevents the tool from cutting too deeply, acting on the *copying principle*, taking off the inequalities and giving the general average of the surface on which the sole rests.

THE PRINCIPLE OF THE SCREW.

92. The screw is a combination of the lever with the inclined plane, and the power gained depends both on the length of the arm of the working lever, and on the inclination of the thread or inclined plane which supports the weight to be raised. We must premise the following definitions :—

Def. 1. If a horizontal line A P which always passes

through a fixed vertical line A B, be made to revolve uniformly in one direction, and at the same time to ascend or descend with a uniform velocity, it will trace out a *screw surface.* This may be readily understood by threading a number of strips of wood on a vertical axis and giving them the successive positions of the lines from B R to A P.

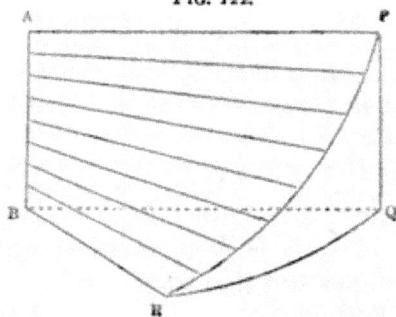

FIG. 122.

Def. 2. The points of intersection of this generating line with any circular cylinder whose axis is A B, will form a *screw-thread,* P R, upon the surface of the cylinder.

Def. 3. The *pitch of a screw* is the space along A B through which the generating line moves during one revolution.

Def. 4. The line A B is the *length* of the surface A B P R.

Def. 5. The angle P R Q is the *angle* of the screw-thread.

In the diagram, A P is shown as describing a *right-handed* screw; if it revolved in the opposite direction during its descent, it would describe a *left-handed* screw.

The screw-thread used in machinery is a projecting rim of a certain definite form, running round the cylinder and obeying the same geometrical law as the ideal thread we have just described.

The screw-thread commonly works either in a nut or against the teeth of a wheel. A nut is an exact copy of the screw, having a recessed part into which the projecting rim accurately fits. The pressure exerted by the screw is felt as a reaction between the surfaces which are in contact. It is at each point perpendicular to those surfaces when friction is neglected, and we shall therefore examine the principle of the screw by conceiving the case of an ideal spiral thread running round the surface of a vertical cylinder and supporting a weight distributed along a definite portion of it.

The lever handle which works the screw will be an arm or line standing at right angles to the axis, it will produce a horizontal pressure wherever the weight acts, and the case will be that of an inclined plane where the power acts in a line parallel to the base of the plane.

93. *Prop.* To find the relation between P and W in the screw.

Let a power P, acting on arm C A, support a weight W by means of a screw thread working in a nut. Conceive that the thread is a geometrical spiral line running round the cylinder and that friction does not act.

Let w_1 be that portion of W which is supported by a portion of the power P, viz. p_1. Then, by Art. 33, Cor. 1, we have $p_1 = w_1 \tan a$, where a = angle of the thread.

Similarly, if $w_2, w_3, \ldots$ be the portions of W supported by the forces $p_2\, p_3 \ldots$ we have in each case $p_2 = w_2 \tan a$, $p_3 = w_3 \tan a$, and so on. Whence, by addition,

$$p_1 + p_2 + p_3 + \ldots = (w_1 + w_2 + w_3 + \ldots) \tan a.$$

Let r be the radius of the cylinder on which the thread is traced, then, by taking moments, we obtain

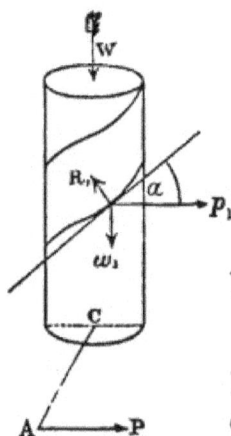

FIG. 123.

$$\text{P} \times \text{C A} = (p_1 + p_2 + p_3 + \ldots) r.$$
$$\text{Also W} = w_1 + w_2 + w_3 + \ldots$$

∴ the condition of equilibrium becomes

$$\frac{\text{P} \times \text{C A}}{r} = \text{W} \tan a,$$

or P × C A = W r tan a.

If we multiply both sides of this equality by $2\,\pi$, we shall have

$$\text{P} \times 2\,\pi\,\text{C A} = \text{W} \times 2\,\pi\,r \times \tan a.$$

Now $2\,\pi\,\text{C A}$ is the circumference of the circle described by the point of application of P, and $2\,\pi\,r \tan a$ is the pitch of the screw-thread. This latter point will be understood on referring to fig. 122, where $PQ = RQ \tan PRQ = RQ \tan \alpha$; it being evident that when RQ is made equal to $2\pi r$, or to the circumference of the cylinder on which the thread is traced, PQ becomes the pitch of the screw. Hence the condition of equilibrium takes the following form.

P (circumf. of circle described by P) $=$ w (pitch of screw).

This result might also have been obtained at once from the principle of work.

The work done by P in one revolution of the screw is $P \times 2\pi CA$, and the work done on w in the same time is w × pitch of screw-thread. But these are equal,

$$\therefore P \times 2\pi CA = W \times \text{pitch of screw-thread.}$$

Ex. Take the case of a screw actuating the break of a railway carriage. Let the radius of the hand lever wheel which turns the screw be 7 inches, and the pitch of the screw be $\frac{1}{2}$ an inch. Find the advantage gained by the combination.

Here the circumference of the circle described by P

$$= 2\pi \times 7 = 2 \times \frac{22}{7} \times 7 = 44 \text{ inches nearly.}$$

Also the pitch of the screw is $\frac{1}{2}$ an inch, therefore the resistance moves through $\frac{1}{2}$ an inch while the power moves through 44 inches.

Hence $P \times 44 = W \times \frac{1}{2}$, or $W = 88\,P$, and the advantage gained is 88 to 1.

94. To find P to w when the screw is rough, let $CA = a$, $\mu = \tan\theta =$ the coefficient of friction. Then

$$p_1 \cos\alpha - w_1 \sin\alpha - \mu R_1 = 0,$$
$$p_1 \sin\alpha + w_1 \cos\alpha - R_1 = 0,$$

whence we deduce $p_1 = w_1 \dfrac{\sin\alpha + \mu\cos\alpha}{\cos\alpha - \mu\sin\alpha} = w_1 \tan(\alpha + \theta)$.

By adding together $p_1 p_2 \ldots$ we obtain finally

$$P = \frac{W r}{a} \cdot \tan(\alpha + \theta).$$

THE SCREW-THREAD.

95. The *screw-thread* used in machinery is a projecting rim of a certain definite form, running round a cylinder, and following the same geometrical law as that referred to in the definition. The two principal forms used by engineers

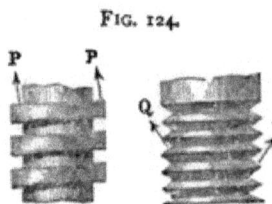

FIG. 124.

are the square and the v thread, shown in the sketch. In speaking of the pitches of such screws, it is the practice to count the number of ridges which occur in an inch of length of the screw bolt, and to esti-mate the pitch by the number of such ridges. Thus a screw of $\frac{1}{8}$-inch pitch is called a screw with eight threads to the inch.

To Sir Joseph Whitworth we owe the introduction of an uniform system of angular screw-threads. The *Whitworth thread* was selected after a careful comparison of the threads in use by engineers, and is everywhere adopted. The angle of the v thread is 55°, but the top and bottom of the edges are rounded off one-sixth part. (*See* 'Workshop Appliances,' p. 103.)

There are three essential characters belonging to a screw-thread, viz., its *pitch, depth*, and *form ;* and three principal conditions required in a screw when completed, viz., *power, strength*, and *durability*. In considering how these several qualities are related, we observe that

1. the *power* of a screwed bolt depends on the *pitch* and *form* of the thread.

If the screw-thread were an ideal line running round a cylinder, the power would depend solely on the pitch, accor-ding to the relation in *Art.* 93.

$$\text{w} \times \text{pitch} = \text{p} \times 2\pi \times (\text{arm of lever}).$$

If the thread were *square*, we should substitute for the ideal line a small strip of surface, being a portion of the surface shown in Fig. 122, which would present a reaction p to the weight or pressure, everywhere identical in direc-tion with that which occurs in the case of the ideal thread. Hence, if there were no friction, we should lose nothing by the use of a square thread in the place of a line. A square-threaded screw is therefore the most powerful of all, and is employed commonly in screw presses. If the thread were *angular*, the reaction Q which supports the weight or pressure would suffer a second deflection from the direction of the axis of the cylinder, over and above that due to the pitch, by reason of the dipping of the surface of the angular thread, and we should be throwing away part of the force at our disposal in a useless tendency to burst the nut in which the screw works. In this sense, the angular thread is less powerful than the square thread.

2. The *strength* depends on the *form* and *depth*.

This statement is obvious. In a square thread half the material has been cut away, and the resistance to any stripping of the thread must be less than in the case of angular ridges. Again, if we deepen the thread we lessen the cylinder, from which the ridges would be stripped if the screw gave way ; and thus a deep thread weakens a bolt.

3. Finally, the *durability* of a screw-thread depends chiefly on its *depth*, that is, on the amount of bearing surface exposed to wear, and resisting the pressure.

CHAPTER V.

ON THE EQUILIBRIUM AND PRESSURE OF FLUIDS.

96. When the molecules of a substance can be separated and moved among each other by the smallest conceivable effort, the substance is called a fluid. There is no very perfect agreement as to the definition of a fluid ; that given by Professor W. H. Miller of Cambridge is the following :

A fluid is a body which can be divided in any direction, and the parts of which can be moved among one another by any assignable force.

There are two classes of fluids, viz., *liquids* and *gases*.

The conception which everyone forms for himself of the
distinction between liquids and gases will hardly be assisted
by any remarks, but as a matter of fact, if we pour any liquid
such as water into a tumbler, it will lie at the bottom and
will be separated by a distinct surface from the air above it.
The same thing is not true of a gas, for a gas is a substance
which however small a quantity we introduce into an empty
and closed vessel, it will immediately expand so as to fill
the whole vessel, and will exert some amount of pressure
upon the interior surface.

*It is the power of indefinite expansion which distinguishes
a gas.*

Gases are further subdivided into permanent gases and
vapours.   There are some gases which, by the action of external pressure or of cold, or by both actions combined,
may be converted into liquids. Of these steam, or the vapour
of water, is an example familiar to everyone.   The invisible
or true gas constituting the vapour of water is always present
in the air, and is readily deposited as dew on any chilled
body.   Again, carbonic acid is a substance never absent
from the air we breathe, and is called a gas, though it is also
recognised as a vapour.   The liquefaction of carbonic acid
gas has been a favourite experiment with philosophers.   At
a pressure which may be roughly estimated at between 40
and 50 times greater than that of the air around us, carbonic
acid becomes liquid.   The liquid state is not permanent
however, for if we tap the vessel containing the substance
and let out a little of the liquid, a jet of spray flies out which
changes into gas in an instant.   By an artifice the issuing
liquid may be frozen into a solid, and then we can experiment with the intensely cold frozen spray which is the solid
form of carbonic acid.

Some gases have never been liquefied ; of these oxygen,
hydrogen, and nitrogen are examples, and are called perfect
gases.

It will be readily understood that if we look merely to

that mobility of the particles which is deemed the characteristic of a fluid there is a wide difference of behaviour in different substances. The particles of all gases are wonderfully mobile, whereas liquids exhibit great differences in mobility. Thus, new honey or tar are imperfect liquids as compared with water, though enormously superior in liquidity to ice. Yet a river of ice will flow slowly down a gorge in the Alps, and its centre will move more rapidly than the edges just as if it were a river of liquid water pouring along a nearly level bed. When subjected to enormous pressures certain solids behave like liquids, and the flow of solids under pressure has recently formed an interesting subject for discussion and enquiry.

Anyone may satisfy himself that ice will flow round an obstacle by the following experiment. Take a block of ice a few inches square, and suspend two weights, say of 5 lbs. each, at the ends of a piece of copper wire about the thickness of pianoforte wire. Place the wire thus loaded over the block so that the ends hang down, and it will begin slowly to enter the ice, which will close up as it passes so that after half an hour the wire will be embedded at some depth in a solid *uncut* mass of ice. The particles of ice have passed round the wire just as the particles of water would pass round a boat which was being towed along a canal.

A FLUID CANNOT RESIST A CHANGE OF SHAPE.

97. Our next observation is that a liquid differs essentially from a solid in being destitute of the power of sustaining pressure unless it is supported laterally in every direction.

A small cylinder of steel, say $\frac{1}{2}$ an inch in diameter, will support an enormous pressure in the direction of its axis, before it becomes compressed into a flat button, but a small quantity of water could not even take the form of a cylinder, for it would not even sustain its own weight unless aided by external support laterally.

Mr. Maxwell distinguishes solids from liquids in the

manner pointed out. *Bodies which can sustain a longitudinal pressure, however small that pressure may be, without being supported by lateral pressure, are called solid bodies. Those which cannot do so are called fluids.*

In other words, a fluid is a body incapable of resisting a change of shape.

98. This fact enables us to comprehend the principle of the well-known hydraulic cranes. A quantity of water is pumped into a strong iron cylinder, and is compressed by a plunger loaded with, perhaps, 70 tons. A pipe passes from the cylinder to a distant crane, the water is pressed down by the weight of 70 tons, and cannot sustain the great vertical pressure unless it be supported by a corresponding force in every other direction.

Wherever the water is conveyed this will be true, and wherever an opening is made the water must be supported. Thus the pressure may be transmitted unimpaired to the piston of a crane at a distance of some hundred yards, and by virtue of this property of a liquid it becomes possible to convey to a distant point the potential energy of the raised weight of 70 tons. Every inch that this weight descends measures an amount of work which may be usefully employed in the crane, and energy is carried to a distance just as if it were a material substance. There are 4 miles of pipe laid down in the Victoria docks for the use of the cranes on the different jetties, and the power travels through the entire length of these pipes.

RELATION BETWEEN THE DIRECTION OF THE SURFACE OF A FLUID AND THE FORCES WHICH ACT ON IT.

99. A direct consequence of this mobility of the particles is that *the surface of a fluid at rest must always be perpendicular to the force which acts upon it at every point.*

Let A B represent the surface of a fluid at rest, P a force acting on a molecule D in the direction D P.

Then P may be resolved along the surface and perpen-

dicular to it. Since the molecules of the fluid are perfectly
mobile there is nothing to prevent the resolved part of P in
direction D A from moving D.
The molecule D would assuredly
glide over its neighbours unless P
acted perpendicularly to A B. It
receives no support except from
the reaction of the surrounding

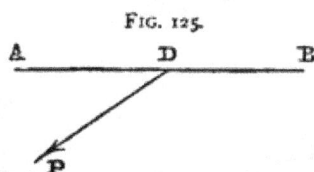

FIG. 125.

molecules, and can only remain at rest in virtue of the
impossibility of penetrating the crowd of particles which are
opposed to it.

THE SURFACE OF MERCURY OR WATER AT REST IS A HORIZONTAL PLANE.

100. If it be true that the surface of a liquid is perpendicular to the force which acts upon it at every point, viz.,
the force of gravity, it must follow that the surface of a small
quantity of liquid such as mercury will be a horizontal plane.
Strictly speaking the surface is curved, but the curvature
cannot be detected because the deviation is infinitesimal for
small areas of surface. In selecting a liquid for experiment
it is evident that the brilliant reflective power of clean mercury eminently fits it for our purpose, when we require a
perfectly smooth and perfectly horizontal mirror. In the
Greenwich Observatory a tray of mercury is placed below
the *transit circle* telescope in such a position that a star
which has been viewed directly may also be seen by reflection
in the surface of the mercury. It is certain that this surface
is a true horizontal plane, wherefore the angle between the
two directions of the optical axis of the telescope will be
twice the angle enclosed between a line pointing from the
eye of the observer to the star, and a second line drawn
through the same point in a true horizontal direction, and
coinciding with the plane swept out by the axis of the
telescope.

In virtue of the same property the astronomer is able to

fix the direction of a true vertical line in his telescope with an accuracy which far surpasses the common observation with a plumb line.

A bowl of mercury is placed directly under the telescope, the cross wires are illuminated, and are observed when the telescope is pointed vertically downwards upon the surface of the mercury. A reflected ray of light coincides with the incident ray only when it is truly perpendicular to the reflecting surface ; and thus a double image of the wires will usually be seen, one by light passing directly from the wires to the eye, the other by light which has been reflected from the mercury ; but if we adjust the telescope so that the two images coincide, we are assured that we have determined a line perpendicular to a horizontal plane, and which is therefore a true vertical line.

THE DIRECTION OF THE PRESSURE OF A FLUID ON A SURFACE.

101. In all cases where we are concerned with forces, the principle that action and reaction are equal and opposite will apply.

Since a fluid acted on by force presents a surface which is everywhere perpendicular to the direction of the force, conversely a surface supporting a fluid which presses upon it will supply a reaction or force everywhere perpendicular to the surface. Whether the pressure produces the surface, or the surface gives rise to the pressure, is immaterial.

Hence it is an established principle, that the pressure of a fluid upon any surface with which it is in contact must be perpendicular to the surface at every point.

This conclusion is exemplified in every possible manner. The pressure of water against the plane surface of a valve is perpendicular to that surface, whatever be the direction of the pipe leading to the valve. We estimate the strain on a cylindrical boiler by supposing that the force of the steam acts on each element of its area in a line perpendicular to that portion, and if the force of the water tends to tear a lock gate off its hinges, it does so by pressing at each point in a line perpendicular to the surface of the gate.

The conception of perfect mobility involves the assumption that there is no friction and no power of cohesion among the particles of a fluid. Granting the possibility of this ideal conception we obtain another definition of a fluid.

Def. *A fluid is a body, the contiguous parts of which act on one another with a pressure which is everywhere perpendicular to the surface which separates those parts.*

EQUAL TRANSMISSION OF FLUID PRESSURE.

102. If the pressure of a fluid on a finite surface in contact with it be perpendicular to that surface, so also the pressure on the surface separating any two sides of a liquid particle within its mass must be perpendicular to that surface. We reason here from the finite to the infinitely minute, because we believe the law to be perfectly general.

Nothing is known about the surfaces which bound a liquid particle, but we assume that the directions in which these surfaces can lie are infinite in number, and that any one of them may determine the direction in which force is being transmitted. Hence a fluid is capable of transmitting force equally in all directions, and further, there is no loss of force in the transmission. Unlike a solid where pressure is only transmitted in the line of its action, the pressure which can be felt at any point in the interior of a mass of fluid is exerted not merely in the line of its original action, but in every other conceivable direction at the same time. The molecules of a liquid can only support each other by direct pressure, they do not cohere, they are destitute of the resisting power due to friction, and they are therefore ready to pass off in a moment in every direction where a path is opened for them, or in which motion is possible.

103. We may here remark that *all fluids are imperfect,* and that whereas a great variety of fluids transmit pressure equally in every direction when at rest, yet they fail to do so when in motion. The truth is, that fluids exhibit an amount of internal friction which is inconsistent with our theoretical conception, and thus the experiments of Mr. Joule, by which the mechanical equivalent of heat was ascertained, were founded on the possibility of increasing the temperature of water or mercury by the heat due to the friction of the molecules. Mr. Joule set a system of paddles in motion within a vessel of water, and rotated the paddles by a descending weight. The water rose in temperature as the weight descended. The heat given to the water found its equivalent in the work done by the descending weight, and the conclusion was that the work required to raise one pound of water from 39° F to 40° F. was equivalent to that done in raising 772 lbs. through one foot.

THE MEASUREMENT OF FLUID PRESSURE.

104. Hitherto we have used the term pressure in its ordinary sense, but it is necessary to point out that the word has a technical meaning, and is used to denote the *pressure in pounds on a square inch of surface.* When we say that the pressure of steam in a boiler is thirty pounds, we mean that the pressure of the confined gas on each square inch of surface of the shell is thirty pounds.

Def. *The pressure of a liquid at a given point is the ratio of the whole pressure on a very small surface, having the point in its centre of gravity, to the area of that surface.*

In other words, if $p\,a$ be the whole pressure on the small surface whose area is (a), the pressure of the fluid at the given point is p.

When the pressure is the same at every point, the pressure on a unit of area is $p \times 1$ or p. That is, the pressure at every point is the whole pressure on a unit of area. Hence the common method of estimating pressures by pounds on the square inch is perfectly correct for uniform pressures.

When the pressure varies, as is the case where water

presses against the gate of a lock, we should estimate the pressure at any point by the force *which would be exerted on a square inch* of surface having the point in its centre, if the intensity of the pressure were uniform throughout the area and equal to that at the point considered.

THE IMAGINARY SOLIDIFYING OF A PORTION OF A FLUID.

105. When a liquid changes into a solid, as when water changes into ice, or when molten wax solidifies, some change of volume is observed. Thus water expands in the act of freezing, while almost every other liquid substance contracts, and in every case the act of solidifying is attended with an alteration of volume. Nevertheless, it is an extremely convenient hypothesis, when reasoning upon fluid equilibrium, to suppose that a portion of the fluid solidifies without any change in bulk. We argue as follows : the fluid is at rest, its particles have no motion, and it is evident that we shall neither alter the pressure nor disturb the equilibrium of the surrounding fluid if we imagine that the power of motion is taken away from a group of particles, or that they are transformed into a solid mass. We shall soon comprehend the value of this hypothesis by which it becomes possible to apply, in the case of fluids, those laws of equilibrium which are known to be applicable to solid bodies.

106. This principle may be applied with advantage in proving that fluids press equally in all directions.

Let the fluid contained within the prism A *b c*, in the interior of a fluid at rest, become solid. If the sides of this prism be indefinitely diminished, we may disregard any external forces acting upon it, such as its weight, and suppose it to be kept at rest by the pressures on its ends and sides.[*]

FIG. 126.

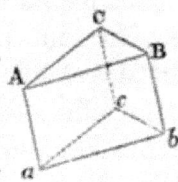

---

[*] The weight is proportional to Aa^3, and the pressures are proportional to Aa^2, hence the former may be disregarded in comparison with the latter when Aa is indefinitely diminished.

Conceive further that the plane ends of the prism are perpendicular to its sides, these pressures will be separately in equilibrium. Since the pressures on A b, A c, C b, are in equilibrium and perpendicular to the sides A B, A C, C B, of the triangle A B C, they are proportional to those sides.

Let p.A b, q.A c, be the pressures on the sides Ab, A c respectively.

$$\text{Then } \frac{p.\text{A} b}{q.\text{A} c} = \frac{\text{A B}}{\text{A C}}. \text{ But } \frac{\text{A} b}{\text{A} c} = \frac{\text{A B}}{\text{A C}}, \therefore p = q.$$

But p, q, measure the pressures of the fluid at A in directions perpendicular to A b, A c respectively, which are chosen at pleasure; hence fluids press equally in all directions.

DEFINITIONS RELATING TO FLUIDS.

107. We pass on to some definitions relating to fluids.

Def. The *mass of a fluid* is estimated exactly as in the case of a solid body.

Def. The *density of a fluid* is the number of units of mass contained in a unit of volume.

A cubic foot is commonly chosen as the unit of volume and a pound may be taken as the unit of mass. One cubic foot of water weighs 62·5 pounds, hence the density of water is 62½ pounds to the cubic foot. This value is approximate, the more accurate statement being that a cubic foot of water at 39°·4 F. weighs 62·425 lbs.

This definition applies to uniform fluids. It may be well to point out that problems are often set in mathematical treatises where the density is assumed to vary according to some definite law. It is scarcely possible to deal with cases of this kind, or with any others involving variable magnitudes, in an elementary treatise.

Practically the fluids with which we are concerned may be regarded as uniform in density; thus water is so little affected by external pressure that it was long believed to be absolutely incompressible; and in truth the application of a pressure of one atmosphere, or about 15 pounds on the square inch, would compress a mass of water somewhat less than $\frac{1}{20.000}$th of its volume. In the hydrostatic press we might get as far as 500 atmospheres, and the compression of the water would then be less than $\frac{1}{40}$th part of its bulk.

The density of air changes with every fluctuation in pressure or temperature. Accordingly the pressure of the atmosphere alters sensibly as we ascend a mountain, and the reading of a portable aneroid barometer enables us to form a fair estimate of the height of the ascent. The fluctuations in density which occur in any separated mass of air are commonly so minute that they may be disregarded, and for most purposes the problems which relate to the equilibrium of gases are treated as simply as those relating to the equilibrium of a mass of water.

It is useful to form some idea of the weight of air. One hundred cubic inches of dry air at 60° F., and 30 inches pressure, weighs 31·0117 grs., whence 1 cubic foot of air weighs ·07638 lbs., and 13 cubic feet of air weigh 1 lb.

The term *specific gravity* is used in a technical sense, which should now be understood. The word *gravity* signifies weight and *specific* means peculiar to the substance. Thus a cubic inch of platinum has a specific weight depending on its being platinum, and different from that of a cubic inch of tin, which can also be specified. In like manner we speak of *specific heat,* or *specific inductive capacity for electricity.*

Def. The *specific gravity of a solid or liquid* substance is the ratio of its density to that of water, or, in other words, it is the ratio of the weight of a given volume of the substance to that of an equal volume of water.

The water is distilled, and the temperature is 60° F.

Def. The *specific gravity of a gas* is the ratio of the weight of a given volume of the gas to that of an equal volume of air at the same temperature and pressure.

The standard temperature is 60° F., and the pressure is that capable of sustaining 30 inches of mercury.

108. In analysis we represent the density of a substance by the Greek letter ρ, which stands for the mass of a unit of volume. For example, if the substance were platinum, ρ would represent the number by which the mass of a cubic inch of distilled water at 60° F. should be multiplied in order to obtain the mass of a cubic inch of platinum.

The specific gravity of the substance is the weight of the matter contained in a unit of volume, and is represented by $g\rho$, in gravitation measure. Thus we have the relations :

Specific gravity $= g\rho$.

Mass of v units $= \rho$ v.

Weight of v units $= g\rho$ v $=$ v (specific gravity).

Ex. 1. Given that a pint of water weighs 20 oz., and that the specific gravity of proof spirit is 0·916 : what fraction of a quart of proof spirit will weigh 30 oz ?   *Ans.* $\frac{375}{458}$.   (Science Exam. 1872.)

Ex. 2. A cup when empty weighs 6 oz. ; when full of water it weighs 16 oz. ; when full of petroleum it weighs 14$\frac{3}{4}$. What is the specific gravity of petroleum? *Ans.* 0.875. (Science Exam. 1871.)

LAW OF INCREASE OF PRESSURE BELOW THE SURFACE OF
WATER.

109. *Prop.* To find the law of the increase of pressure at increasing depths below the surface of a liquid when at rest and acted on by gravity.

Let C A D be the surface of the liquid at rest, and conceive that the portion contained within a slender prism A B,

FIG. 127.

having *vertical* sides and a *horizontal* base, is converted into a solid. This will neither

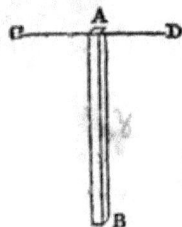

alter the pressure, nor disturb the equilibrium of the surrounding fluid.

The prism A B is now a solid at rest under certain forces, viz., its own weight, and the pressures on its ends and sides.

But the surface at A is horizontal because gravity is the only force acting ; also the surface at B is horizontal, by hypothesis, and the sides of the prism are vertical. Wherefore the pressures on the ends and the weight of the prism are vertical forces, while the fluid presssures on the sides are horizontal forces, and each group is separately in equilibrium.

∴ pressure on end B = weight of prism + pressure on end A.

Let p be the pressure at B, a the area of a horizontal section of the prism, A B $= z$, and w the weight of a cubic inch of the liquid. Then

$p a$ = weight of the prism A B + pressure on end A,
$= a z w$ + pressure on end A.

For small depths, we may take inches as units; for greater depths, we may measure in feet or fathoms, also it is usual to disregard the pressure at A, as we are not sensible of the pressure of the atmosphere under ordinary circumstances.

Hence $p = z w$, where z is the depth in inches.

The law of pressure therefore is, that the pressure varies as the depth, when the surface pressure is neglected, and that its increase is in exact proportion to the increase of vertical depth.

Note.—This theorem is true for liquids but not for gases. It proceeds on the hypothesis that the prism A B is of uniform density throughout, which would be untrue if the fluid were compressible, and became more dense at increasing depths.

Ex. 1. At what depth in water does p become 15 lbs.?

Here $144 p = z' \times 62·5$; ∴ $z' = \dfrac{15 \times 144}{62·5} = 34·56$ feet.

Ex. 2. At what depth in mercury does p become 15 lbs. ?

Here $p = z \times$ weight of a cubic inch of mercury, *i.e.*, $\dfrac{3429·5}{7000}$ lbs.,

therefore $z = \dfrac{15 \times 7000}{3429·5} = 30·6$ inches.

Ex. 3. The pressure of water used for working hydraulic cranes is 700 lbs. on the square inch. To what *head* or vertical depth does this correspond ?

∵ $700 = 15 \times \dfrac{140}{3}$, the '*head*' is $\dfrac{140}{3} \times 34·56$ or $1612·8$ feet.

The law of the increase of liquid pressure may be expressed analytically, for let ρ be the density of the fluid, and m the pressure at A, then

$$p\,a = g\rho a z + m\,a, \text{ or } p = g\rho z + m.$$

110. Since the pressures are equal when the depths are equal, it follows that surfaces of equal pressure are also surfaces of equal depth. But the surface of the liquid has been shown to be a horizontal plane (*See Art.* 100), wherefore a surface of equal pressure is everywhere at the same depth below a horizontal plane, and is itself a horizontal plane.

For a like reason, when any number of vessels containing the same liquid are in communication, the liquid stands at the same height in each vessel.

Again, the common surface of two fluids that do not mix must be a surface of equal pressure, and the upper surface of the lighter fluid is a horizontal plane, therefore the common surface of two fluids that do not mix is a horizontal plane.

Prop. *When two liquids that do not mix rest against one another in a bent tube, the altitudes of their surfaces above the horizontal plane in which they meet are inversely as their densities.*

Let x, y, represent the altitudes above the common surface of the liquids whose densities are ρ, σ respectively.

Since the common surface is one of equal pressure,

we have　　　　$g\rho x = g\sigma y.$

$$\therefore\ x : y = \sigma : \rho.$$

111. *Prop.* To find the pressure of a liquid on any surface.

Conceive that the surface is divided into an indefinite number of minute portions $s, s', s'', \ldots$ at the respective depths $z, z', z'', \ldots$ below the level of the liquid.

Let G be the centre of gravity of the surface, $\bar{z}$ its depth below the level of the liquid, $p, p', p'' \ldots$ the pressures on $s, s', s'' \ldots$ respectively, and w the weight of a cubic inch of the liquid.

Then $p = szw$, $p' = s'z'w$, and so on.

$$\therefore p + p' + p'' + \ldots = w(sz + s'z' + s''z'' + \ldots)$$
$$= w\bar{z} \times (\text{area of surface}).$$

$\therefore$ Whole pressure on the surface $= w\bar{z}$ (area of surface).

That is, *the pressure on the surface is the weight of a column of liquid whose base is the area pressed, and altitude the depth of the centre of gravity of the surface below the level of the liquid.*

This proposition is general and applies indifferently to curved or plane surfaces.

Ex. 1. Find the pressure on the internal surface of a sphere when filled with water.

Let a = radius of sphere, w = weight of a cubic inch of water, *i.e.* 252·7 grains, then pressure on surface $= 4\pi a^2 \times aw = 4\pi a^3 \times w$
$$= 3 \times \text{weight of the water}.$$

Ex. 2. A hemispherical cup is filled with water and placed with its base vertical ; compare the pressures on the curved and plane surfaces.

Here pressure on curved surface $= 2\pi a^2 \times aw = 2\pi a^3 w$.

Pressure on plane surface $= \pi a^2 \times aw = \pi a^3 w$.

Hence the pressures are as $1 : 2$.

112. The last example leads us to distinguish between the total pressure of a fluid on a curved surface, and that portion of it which is perpendicular to any given plane. The pressure on the vertical plane side of the hemispherical cup might be obtained by adding up the horizontal components of the actual pressures on each small element of the surface. The pressure so obtained is called the *resultant horizontal pressure* of the liquid on the surface, and is equal to the liquid pressure on the base, for otherwise the cup would have a tendency to move in a horizontal direction, which is contrary to experience.

113. Prop. *The pressure on the base of a vessel containing any liquid is independent of the form of the vessel.*

This appears from *Art.* 110, and is also a direct consequence of the fact that the pressure on the base of a vessel depends simply on the area of the base and the depth of its centre of gravity below the surface of the liquid.

Ex. 1. Suppose a right cone to be filled with water and placed on a horizontal plane.

Let b = altitude of cone, a = radius of base, then pressure on base = $\pi a^2 b w$ = 3 times the weight of the enclosed water.

Ex. 2. Let a circular right cylinder be filled with water and placed on a horizontal plane. Find the pressure on its base. ·

Let b = altitude of cylinder, a = radius of base, then pressure on base = $\pi a^2 b w$ = weight of the enclosed water.

We account for the results in these two cases by attributing the increased relative pressure as compared with the weight, in the former of these two instances, to the reaction of the oblique surfaces which form the sides of the vessel.

The pressure on the curved surface of the cylinder is entirely horizontal, and does not react upon the base in any way; whereas, the pressure on the curved surface of the cone consists of an assemblage of forces whose vertical components all point downwards and react upon the base.

The fact can be verified by experimenting with vessels of water, of various forms, the base of the vessel being the surface of mercury in a siphon tube, and the pressure being noted by observing the elevation of the column of mercury in one leg of the siphon.

In the following examples, the pressure of the atmosphere is disregarded.

Ex. 1. A closed cylindrical vessel, the radius of whose base is 8 in. is filled with water, and placed with its axis horizontal; find the pressure of the water against one end. (A cubic inch of water weighs 252·5 grains.) (Science Exam. 1873.)

Ex. 2. The depth of water in a vessel is 10 feet, the base of the vessel is a square with a side $1\frac{1}{2}$ feet long; what is the pressure on it?

Ex. 3. A square A B C D is immersed in a liquid at some depth, and turns about the horizontal side A B as an axis. When the square is in the lowest position, hanging vertically, the pressure on its face is twice what it would be if the square were rotated about A B so as to take its highest position. Find the depth of A B. (Science Exam. 1872.)

Ex. 4. The pressure of a liquid on a square is $\frac{1}{4}$ the weight of a cube of the liquid, whose edge is equal to a side of the square. If one edge of the square be in the surface of the fluid, what is the inclination of the square to the horizon? (Science Exam. 1871.)

Ex. 5. Find the pressure on a circle 6 inches in diameter when immersed at a depth of one mile below the surface of the sea. (A cubic foot of sea-water weighs 64 lbs.)

Ex. 6. The depth of a dock gate is 32 feet, and its breadth is 35 feet; find the pressure on it when the water rises to a height of 20 feet on the outside.

THE CORNISH DOUBLE-BEAT CROWN VALVE.

114. In order to make the meaning of resultant fluid pressure more clear, we refer to the *Cornish crown valve*, a valuable form of valve adopted originally in the Cornish pumping-engines, and now commonly used both as a steam and hydraulic valve.

The principle being the same in either case, it may be convenient to illustrate our proposition by means of an hydraulic valve.

FIG. 128.

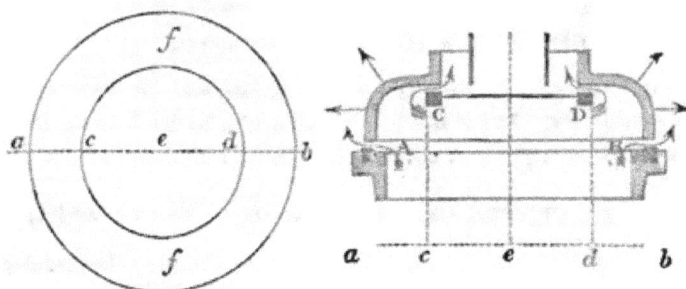

Here a crown-shaped cover, shown in section in the diagram, rests upon two fixed concentric circular seats at C, D, A, B. When the cover rests on its seat, the water cannot pass from below upwards, but as soon as it is raised the water escapes at the two annular openings formed at C, D and A, B.

The only moveable thing being the crown, we will examine its behaviour under the pressure of the water.

Now the fluid pressure acts perpendicularly at every point of the curved surface, and its action on each element of the valve may be resolved in a vertical and horizontal direction. The sum of all the vertical components, *or the resultant vertical pressure*, is equal to the pressure on the annulus formed by subtracting the circle of radius *e c* from that of radius *e a*, the lines *e c*, *e a* being the respective radii of the outer edge of the upper valve seat and the inner edge of the lower one. *The resultant of all the horizontal components is zero.*

1. Let *e c* = *e a*, the vertical pressure upwards is zero, and the valve is a perfectly *balanced* or *equilibrium valve.* In other words, the water pressure has no tendency either to open or close the valve. It may be opened or closed by any force which suffices to overcome its weight and the friction of the working parts.

2. Let *e c* be less than *e a*, the resultant vertical pressure will be that on the annulus *f f.* The direction of this pressure acts upwards, and the valve will open as soon as the pressure on the annulus exceeds its effective weight.

3. Let *e c* be greater than *e a*, the valve will be held down on its seat by the pressure on the annulus.

Ex. Let *e c* = 3¼ inches, *e a* = 5 inches, and let the weight of the valve be 68 lbs. What head of water could be held back by such a valve before the pressure could cause it to lift? (Science Exam. 1870.)

THE CENTRE OF PRESSURE OF A PLANE AREA.

11 . *The centre of pressure of a plane area immersed in a liquid is that point in which the resultant pressure of the liquid acts.* It is lower than the centre of gravity, because it is the centre of a system of parallel forces which increase as the depth increases.

This being an example of the action of variable forces, we do not attempt a general solution, but merely show how to find the centre of pressure when the area immersed is a plane rectangle, having one side in the surface of the liquid. We disregard the pressure of the atmosphere because the effects produced by it are commonly balanced.

Let F E be such a rectangle, C Q the edge of a narrow strip of the surface, made by two vertical planes. Draw

Q R horizontal and equal to c Q, join c R ; **also** from any point p in c R, draw n P parallel to Q R, and **complete** the rectangle P *n*, whose vertical depth is very small. Since the pressure

FIG. 129.

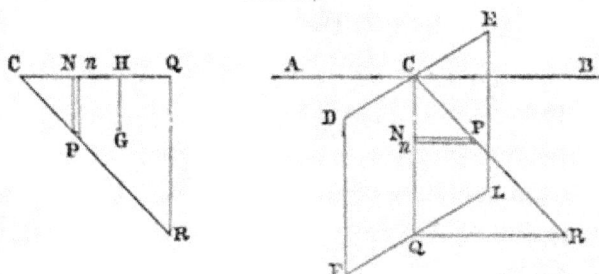

varies as the depth, the pressure at Q is proportional to c Q, and therefore to Q R; so likewise the pressure at N is proportional to N P.

By reasoning exactly similar to that employed in page 30, **we** can show that the area of the triangle c Q R represents the whole amount of fluid pressure on the rectangular strip whose edge is c Q. In other words, if c Q were a horizontal line, and the triangle c Q R were made up of a series of heavy wires hanging upon c Q, we should create a mechanical equivalent for the actual fluid pressure on the strip. Let G be the centre of gravity of the triangle c Q R, G H a vertical line through G, then c H $= \frac{2}{3}$ c Q. Now the weight of c Q R acts through H, therefore also the resultant of the fluid pressure on the strip acts through H, and the depth of the centre of pressure is $\frac{2}{3}$ the depth of the immersed edge D F. It is also evident that the centre of pressure lies in a vertical line bisecting D E.

Ex. **1.** Find the centre of pressure of a triangle whose base is horizontal and vertex in the surface of the fluid. (Science Exam. 1870.)

Ex. **2.** Find the centre of pressure of a right-angled triangle A C B, having the side c B in the surface.

Bisect c B in E, also divide A c in the point **F** such that c F $= 2$ A F ; join c E and F B intersecting in Q, which will be the centre of pressure of the triangle, and it is easy to show that the depth of Q is $\frac{1}{2}$ A C.

Ex. **3.** The breadth of a water passage closed by a pair **of** gates is 10 feet, and its depth is 6 feet. The hinges are placed at one foot from the top and bottom ; find the strain on the lower hinge when the water rises to the top of the gates on one side.   *Ans.* 4,218¾ lbs.

In the last example the atmospheric pressure could not enter, for its direct pressure on one side of the area balances the transmitted action on the other side.

THE CONDITIONS OF EQUILIBRIUM OF A FLOATING BODY.

116. The power of floating in liquids, or even in gases, which many substances possess is full of interest, because we recognise in it the evidence of a natural law.   The discovery of the law is due to Archimedes, and the statement of it is the following :—

1. *When a solid floats in a fluid the weight of the solid is equal to the weight of the fluid displaced.*

2. *The straight line which joins the centres of gravity of the solid and fluid displaced is vertical.*

There are some truths in mechanics which we perceive by intuition, and this is one of them.   When a ship floats on the sea we know that every portion of water is pulled downwards by gravity, and that the ship itself is pulled down in like manner.   We know also that the surface of water at rest is a level plane.   The ship must therefore sink till it has reached such a position that the level of the sea is again made perfect, not by a mass of water, but a vessel equal to it in weight.

In order to establish the law let us suppose that a portion of water, as D E F, contained within a mass of water at rest, becomes solidified.   (*See Art.* 105.)   Thus, D E F is a

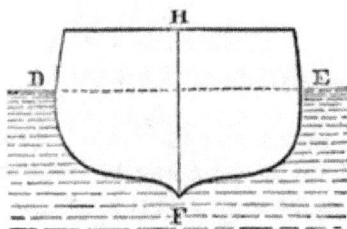

FIG. 130.

solid at rest under

1. Its own weight acting downwards through the centre of gravity.

2. The pressure of the surrounding liquid.

These are the only forces acting, and the body is at rest ;

therefore the resultant pressure of the fluid must be a force equal to the weight of D E F and acting upwards in a vertical line through the centre of gravity of D E F.

Conceive now that a ship H F, floating in the water, displaces the same volume D E F. The weight of the ship will act downwards in a vertical through its centre of gravity ; the pressure of the water will be the same as before ; also the forces in action must be equal and opposite, and therefore the weight of the ship must be equal to the weight of the water displaced, and the line joining the centres of gravity of the solid and water displaced must be vertical.

This law holds good in the case of any solid, whether wholly or partly immersed in a liquid or gas ; *the weight lost is always equal to the weight of the fluid displaced.*

117. *Prop.* To determine whether the equilibrium of the floating body is stable or unstable.

The general condition will be understood without any difficulty. The drawing shows a section of the vessel H F when heeling over. Let G be the centre of gravity of the vessel, and Q that of the water displaced in the new position, then the weight of the ship is a vertical force W acting downwards through G, while the pressure of the displaced water is a vertical force P acting upwards through Q ; these forces balance, for otherwise the vessel would rise or sink, and they produce a couple whose arm is the perpendicular distance between the lines G W and Q P. Let Q P meet F G H in M, then it is clear that so long as M lies above G the tendency of the water pressure will be to restore F H to the vertical direction, whereas if M lies below G, the couple will tend to deflect F H still further, and the vessel will fall over. Hence the danger of taking the whole cargo out of a vessel

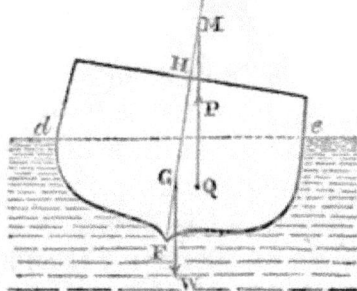

FIG. 131.

without putting in ballast at the same time, or the risk of upsetting when five or six people stand up at once in a small boat. The equilibrium is stable or unstable according as M lies above or below G.

Note. If the vessel were inclined through a very small angle the vertical Q M would meet F G H in a definite point, known as the *metacentre*. The position of this point is ascertained by the aid of the calculus.

118. We have not space to follow this law through its varied applications, and shall merely examine the principle of the methods adopted for finding specific gravities. The student will find a description of the apparatus used for this purpose in many books on chemistry.

THE METHOD OF FINDING SPECIFIC GRAVITIES.

In determining the specific gravity of a solid, the object is to compare its weight with that of the water it displaces. The question may be very approximately solved by neglecting the weight of the air displaced, and the apparatus is a balance of precision, the solid being attached to the bottom of one scale-pan by a fine hair when its apparent weight in water is taken.

Let w be the weight of the solid in air, x its apparent weight in water. Then

weight of solid — weight of water displaced by it $= x$,

$\therefore$ weight of water displaced by solid $= w - x$.

$$\text{Therefore specific gravity of solid} = \frac{w}{w - x}.$$

We have here supposed that the solid is heavier than water, and will sink in it. If it be lighter, we must attach a sinker, so as to make the compound body heavier than water, and proceed as follows.

Let w be the weight of the solid in air, x the weight of the sinker in water, y the apparent weight of the body and sinker in water. Then

weight of solid + weight of sinker — weight of water displaced by sinker — weight of water displaced by solid = Y.
  or w + x — weight of water displaced by solid = Y.

  ∴ weight of water displaced by solid = w + x — Y.

$$\therefore \text{ specific gravity of solid} = \frac{w}{w + x - y}.$$

If the solid be soluble in water, it may nevertheless be
insoluble in some other liquid of known specific gravity,
such as alcohol, or oil of turpentine, &c., and its weight can
be compared with that of the liquid displaced, and therefore
with the weight of an equal bulk of water.

If regard be had to the weight of the air displaced by the
solid, let U represent it ; then the weight of the solid
is w + U, which must be written for w in the preceding
formulæ.

The specific gravities of two liquids may be compared by
weighing equal volumes of each, the specific gravities being
in direct proportion to the weights so ascertained. Or they
may be compared by means of an hydrometer which, in
some form or other, is a hollow vessel, weighted so that it
will float upright, and having a graduated stem which indicates the depth to which it sinks and therefore the volume
of the liquid displaced by it. The specific gravities of the
liquids are inversely as these volumes.

Ex. 1. A body weighs 2,300 grs. in air, 1,100 grs. in water, and
1,300 grs. in spirit ; what is the specific gravity of the spirit ?
                                                    (Science Exam. 1873.)

Ex. 2. A vessel contains mercury (sp. gr. 13·6) in which floats a
cube of iron (sp. gr. 7·2) ; water is poured into the vessel until the cube
is completely covered ; find what portion of the cube is below the surface
of the mercury.   (Science Exam. 1873.)

Ex. 3. The area of the section of a ship made by the plane of the
water is 1,000 square feet. What weight will make her sink 4 inches
lower? (sp. gr. of salt water is 1·026.)   (Science Exam. 1872.)

Ex. 4. A pint of water weighs 20 oz., and the sp. gr. of proof spirit
is ·916 ; what fraction of a quart of proof spirit will weigh 30 oz. ?
                                                    (Science Exam. 1872.)

O

Ex. 5. A piece of cork weighing 1 oz. is fastened to a sinker weighing 3·5 oz. It is found that they will just sink when placed in water. The sp. gr. of cork being 0·25, what is the specific gravity of the sinker? (Science Exam. 1871.)

Ex. 6. A piece of wood weighs 4 lbs. in air, and a piece of lead weighs 4 lbs. in water. The lead and wood together weigh 3 lbs. in water. Find the sp. gr. of the wood. *Ans.* 0·8.

Ex. 7. A body immersed in water is balanced by a weight P, to which it is attached by a string passing over a fixed pulley. When half immersed it is balanced in the same way by a weight 2 P. Find the sp. gr. of the body. *Ans.* $\dfrac{3}{2}$

Ex. 8. A numerous class of questions set in examination papers refer to the discovery of the weight of gold in a mass of gold and quartz. The original problem of finding out the quantity of silver with which the crown of Hiero, king of Syracuse, was adulterated, is stated to have been solved by Archimedes. Let the question be to find the weight of gold in a compound mass of gold and quartz.

Let x, y be the weights of gold and quartz in the mass ; m, n, r the specific gravities of gold, quartz, and of the compound mass, w the weight of the specimen.

$$\text{Then} \quad \text{w} = x + y \ . \ . \ . \ (1).$$

Also volume of specimen = volume of gold + volume of quartz.

$$\therefore \ \frac{\text{w}}{r} = \frac{x}{m} + \frac{y}{n} \ . \ . \ . \ (2).$$

From which two equations x and y can be found.

Ex. 9. A diamond ring weighs 65 grs. in air, and 60 in water ; find the weight of the diamond, the sp. gr. of gold being 17½. and that of the diamond 3½. *Ans.* $\dfrac{55}{8}$ grs.

Ex. 10. The crown of Hiero, with equal weights of gold and silver, were all weighed in water. The crown lost $\frac{1}{14}$ of its weight, the gold lost $\frac{4}{77}$ of its weight, and the silver lost $\frac{2}{21}$ of its weight. Prove that the gold and silver were mixed in the proportion of 11 : 9.

CHAPTER VI.

ON THE EQUILIBRIUM AND PRESSURE OF GASES.

119. WE have now to treat more particularly of those fluids called *gases* which differ from liquids in being capable of indefinite expansion. The theory which accounts for this expansion has been developed by Clausius and Maxwell, and is stated in the treatise on heat. It is only possible to allude to it here, but the student is asked to consider the following points of distinction between solids, liquids, and gases, as preparatory to an observation on the so-called *kinetic theory* of gases.

1. The molecules of a solid can be moved through very minute spaces, but do not pass to a sensible distance from their original position. Such a movement only takes place under the action of force, it consumes work, and the return or rebound of the particles to their normal positions, after displacement, indicates a property termed *elasticity*.

2. The molecules of a liquid are mobile, and it is proved by experiments on diffusion that the molecules of a liquid can move to sensible distances from their normal positions without any apparent cause. In this respect liquids present a striking contrast to solids.

Experiment. A tall glass jar is filled for about ⅔rds of its length with a blue infusion of litmus in water, and some oil of vitriol is cautiously poured in by a long funnel, so as to lie below the water. After two or three days the heavier acid will rise through the water, and its diffusion will be rendered visible by the red colour imparted to the solution. Here the two liquids intermix throughout the jar, the heavier particles moving upwards, and there is no apparent cause for the motion.

3. The molecules of a gas are mobile, and if a quantity of gas, however small, be introduced into a closed vessel, it will instantly fill the whole and will exert some pressure on

the sides of the vessel. Also it is found that all gases diffuse
into each other.

Experiment. Chlorine gas is thirty-six times as heavy as
hydrogen gas, yet if we fill two glass vessels, one with
chlorine, and the other with hydrogen, and connect them by
a glass tube so that the hydrogen is uppermost, the gases
will diffuse, and after a few hours both vessels will be filled
with equal parts of chlorine and hydrogen.

According to the *kinetic theory*, a gas consists of a great
number of molecules, flying in straight lines, and impinging
like little projectiles, not only on one another, but also on
the sides of the vessel holding the gas. It has been stated
that a quantity of gas, however small, will expand and fill
the whole of a vessel, however large ; and further that it will
exert some pressure on its sides ; also, gases of every kind
will diffuse into each other. The expansion and diffusion
of gases are accounted for at once by the theory of molecular
motion ; and so are the laws of Boyle and Charles, presently
to be examined. The molecules should be pictured to the
mind as endued with velocities somewhat· greater than that
of a rifle bullet, and thereby competent to rush into and fill
an empty space with great rapidity. Also, by continually rebounding from the sides of the vessel and from each other they
keep up an incessant cannonade, and the aggregate of these
minute blows is felt as a sensible pressure on the surface
subjected to them. A bladder partly filled with air looks
shrivelled, but when held before a fire it will become hard and
tense. The heat of the fire has given increased velocity
to the molecules, and has enabled them to do more work.
They discharge themselves with greater impetus against the
inner surface of the bladder and overpower the bombard-
ment from without. Presently their power becomes
weakened by the increase of the area to be supported, and
the bladder ceases to enlarge. As the air cools down the
reverse happens, and the bladder soon shrinks back to its
original dimensions. If it be objected that we ought to

have the power of discerning in some way this motion, we answer that the molecules are so minute that their movement is as invisible as the vibratory motion set up in a solid body by heat. The question for the student is not whether he can render an effect of this kind visible, but whether he can satisfy himself by reasoning and experiment that it really must and does exist. As he advances in research he will perhaps find that the arguments in support of it become more impressive, while the power of resistance is enfeebled.

We pass on to the measurement of gaseous pressure.

THE MEASUREMENT OF ATMOSPHERIC PRESSURE.

120. *Prop.* To measure the pressure of the atmosphere.

The experiment now to be described, was first made by Torricelli in the year 1643. A tube A B, closed at one end, about $\frac{1}{3}$ of an inch in diameter, and more than 31 inches long, is filled completely with clean mercury and inverted in a small vessel of the same liquid. The mercury will sink in the tube and rise in the cup, but will speedily come to rest, when a column about thirty inches in length will remain supported above the horizontal plane touching the surface of the mercury in the basin. Since the pressure is the same at all points in this horizontal plane $a\,b$, the weight of the column of mercury resting on the area B c must be equal to the pressure of the atmosphere on the same area.

Fig. 132.

Let P be the pressure of the atmosphere on an area of one square inch, σ (the Greek letter *s*) the density of mercury, h the altitude B C, *measured in inches*, then P is the weight of a column of mercury, of altitude h, and *section one square inch*, therefore

$$P = g\,\sigma\,h = h \times \text{weight of a cubic inch of mercury.}$$

121. *A barometer* is an instrument constructed on the principle described, and is merely a glass tube filled with clean mercury by the process of boiling, the open end being immersed in a basin of the same liquid; the object of boiling is to exclude all traces of air and moisture from the empty space at the top of the tube. A full account of the barometer will be found in many treatises on Physics.

Note. 1. The standard height is sometimes taken as 29·922 inches of mercury, in which case the pressure of the atmosphere is 14·7 lbs. on the square inch, or 2116·4 lbs. on the square foot.

Note 2. It is well known that air becomes less dense as we ascend a mountain, but we can conceive that it remains incompressible, or *homogeneous*, as it is termed. In that case the height of the *homogeneous atmosphere* would be the height of an incompressible mass of air which pressed with a force of 2116·4 lbs. on the square foot. This may be taken to be 26,214 feet. Mr. Glaisher has ascended in a balloon to a greater height than that of the homogeneous atmosphere, viz. to an altitude of about 7 miles.

Ex. 1. A cubic inch of mercury at 16° weighs 3,429½ grs. nearly, and the barometer stands at 30 inches. Find the atmospheric pressure on the square inch of surface.

$$\text{Here } P = \frac{3,429\frac{1}{2} \times 30}{7,000} \text{ lbs.} = 14\cdot698 \text{ lbs.}$$

With the same data, find the height of a barometer filled with water instead of mercury, (sp. gr. of mercury 13·6). Here height required = 13·6 × 30 inches = 34 feet. In practice the height is often taken as 32 feet.

122. The barometer tube supplies a gauge for measuring the pressure of air or vapour in a partially exhausted vessel, such as the receiver of an air-pump, or the condenser of a steam-engine.

If the tube B A were connected at A with a vessel partially filled with air, we should find that the mercury would still rise to some height C in the tube, and since the pressure at B c is always that of the atmosphere, the following equality would maintain, the pressures being taken on the area of an internal section of the tube, which call K,

(Press. of air in AC) K + weight of B C=(Press. of atmosphere)K

If the pressure of the air in A C increases, B C must diminish, but this equality can never be departed from, and the barometer tube furnishes therefore an excellent gauge for measuring pressures below that of the external air. For great accuracy a gauge tube and a barometer should be placed side by side, and should dip into the same basin of mercury. Since 30 inches of mercury produces a pressure of 15lbs. on the inch in round numbers, we infer that a depression of 2 inches of mercury corresponds to 1lb. of pressure in A c, whereby a scale of inches enables us to read pounds of pressure of the enclosed gas with sufficient accuracy for many purposes.

THE PRINCIPLE OF THE SIPHON.

123. The siphon is a bent tube A B C, open at both ends, and used for drawing off liquids from vessels without any agitation. In order to set it in action the tube is filled with the same liquid as that in the vessel, one end is dipped therein, and the other is held below the plane of surface of the fluid. As regards the column B C, we observe (1), that the pressure within the tube at a point D in the plane of the

Fig. 133.

surface of the liquid is the same as at A, and is equal to the pressure of the atmosphere; (2), that the pressure within the tube at B is greater than it is at D, or greater than the atmospheric pressure. Hence the liquid pressure at B overcomes that of the air from without, whereby the column B c tends to separate at C, and to run out at B. If the altitude

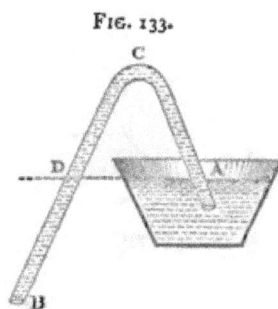

of C above A D be less than the height of the liquid in a barometer tube, the pressure of the air will prevent any separation at C, and will keep up a continuous stream, by forcing the liquid to ascend A C. This will continue so long as B is below the level of the plane A D.

THE SIPHON GAUGE.

124. A siphon gauge is useful for measuring small pressures, such as the pressure of gas supplied to houses, the pressure of the blast of air in a smelting furnace, the pressure of the wind, &c. The liquid used is commonly water. Where the pressures are extreme it may be necessary to employ mercury, which is less sensitive than water in the ratio of 1 to 13·6.

The gauge is a bent tube A C B with parallel legs, partly filled with water. The end A is open, and the end B is in communication with the gas whose pressure is to be ascertained. If the pressure of the gas be *greater* than that of the atmosphere the water will rise in C A and sink in C B, if it be *less* the reverse will happen. Let D E be a horizontal plane touching the surface in B C, and P D the difference of level in the two legs. Suppose, for simplicity, that the internal section of the tube is 1 square inch, then pressure of the gas in B E = weight of P D + pressure of the atmosphere. Hence the excess of pressure of the gas above that of the atmosphere is equal to weight of the liquid column P D.

FIG. 134

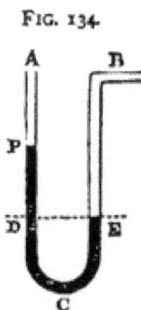

Ex. 1. The pressure of the air supplied by a fan is 6 oz. on the square inch. What column of water will it support in a siphon gauge? *Ans.* 10·39 inches.

125. The siphon gauge may be made more sensitive by an arrangement due to Dr. Wollaston.

Here the ends A and B are two vessels whose sectional

area is considerable as compared with that of the tube. The vessel A is open to the air, while B communicates with the gas under pressure. Pour water into the siphon tube until it is half full, and then fill the tube and both vessels up to a moderate height with oil whose specific gravity is 0·9. If B be opened to the gas under pressure, the water will sink in B E and rise in A D until E D becomes the level of the water in E B, and P Q the level of that in D A. Let K be the area of an internal section of the tube, then

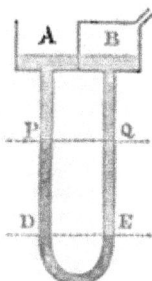

FIG. 135.

(press. in B — press. in A) K = weight of P D — weight of Q E

$$= \tfrac{1}{10} \text{ weight of } P D.$$

Hence P D must be 10 inches in length in order to indicate a difference of pressure in B and A which is really equivalent to 1 inch of water.

The principle of the gauge is now evident: the artifice consists in opposing to P D the weight of an equal column of oil, whereby an apparent inch of water is only an effective $\tfrac{9}{10}$ of an inch. The difference of levels of the oil in A and B has been neglected, but a correction may be introduced for it.

GAS PRESSURE GAUGE.

126. A sensitive pressure gauge may be obtained on a different principle. Let a hollow air-tight cylinder G H be attached to the inside of a cylindrical gasholder D A B C, as shown, and make the volume of G H to that of A C as 1 : *n* or as 1 : 3 in our example. Conceive that the vessel floats as in the right-hand diagram, the level of the water being E R, then the lifting force on the apparatus will be the weight of the water displaced by E H.

Now increase the pressure of the gas in A R till the difference of level of the water within and without the gasholder

is P Q or 1 inch.   The gas will be under a pressure of 1 inch,
and the reading of a siphon gauge would be one inch, but

FIG. 136.

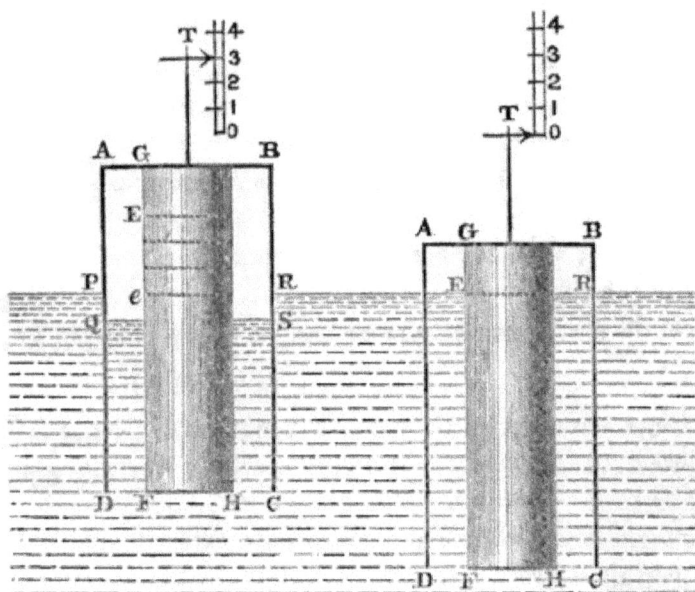

the holder will rise 3 inches, whereby the apparatus is
more sensitive than we might have anticipated in the proportion of 3 to 1.   In order to explain this peculiarity, let
a and $3a$ be the effective areas of G H and A C respectively,
w the weight of a cubic inch of water.   Then

$$\text{volume } P S = 3\,a \times P Q,$$

or the lifting force on the apparatus is increased by

$$3\,a \times P Q \times w,$$

in virtue of the displacement caused by the gas.

Now the work of supporting the apparatus required from
G H is lessened by the amount done otherwise, hence G H
will rise out of the water to a height E e, such that

$$E e \times a \times w = 3\,a \times P Q \times w,$$

$$\therefore E e = 3\, P Q.$$

Or the pointer rises 3 inches for a pressure of 1 inch.   In using this apparatus, the main is connected with the interior of A R, and the fluctuations of pressure can be magnified to any extent required, by simply enlarging the air-tight cylinder

BOURDON'S PRESSURE GAUGE.

127. This instrument has been in use for more than twenty years, and has proved of great value. The circumstances attending its invention show the advantage of reasoning upon observed facts. The worm-pipe of a still had been accidentally flattened, and M. Bourdon endeavoured to restore its circular form by forcing water into it. As the flattened tube became more round it uncoiled itself to a certain extent. It soon became apparent that the action here observed might be applied in the construction of a pressure gauge.

There is a theorem in geometry that if p be a point in a surface, and $a p b$, $c p d$ be two sections of the surface made by planes at right angles to each other, and passing through the normal to the surface at the point p, the sum of the curvatures of the two sections $a b$, $c d$ will be a constant quantity. That is to say, if you turn the whole system

FIG. 137.

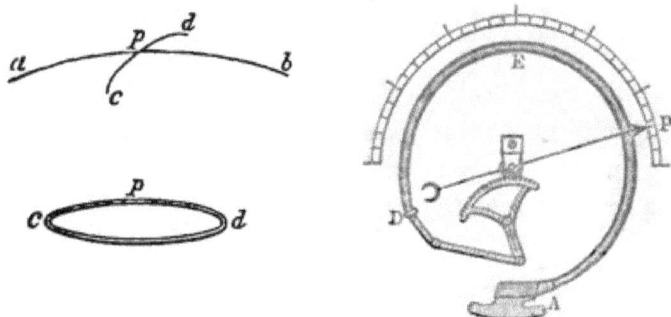

$a b c d$ round the point p, and find that $a b$ becomes more curved, $c d$ will become less curved, and the sum of the two curvatures will remain constant. Take now a flattened circular tube whose cross section is shown at $c p d$, and let $a p b$

represent the curvature of the tube, while *c p b* represents the curvature of the section at right angles to *a p b*. It is evident that when a fluid under pressure is admitted into the flattened tube A B E D, shown in the drawing, it tends to become more circular in the direction *c p d*, and we infer therefore that it must become less curved in the direction *a p b*, or that it will straighten a little. The drawing shows the complete gauge, with the connection of one end of the tube to an index finger P. The principle relied on is, that the end r will move outward as the pressure within the tube increases. It is easy to show that such an action will certainly occur for if we bend a vulcanized indiarubber tube, we shall find that it gets more and more flat, and finally straightens, in the direction across that in which we are bending it. In the gauge, the motion of the end is communicated to the pointer P, and the reading on the scale indicates the pressure of any gas or liquid enclosed within the tube. This form of gauge will measure pressures below that of the atmosphere just as well as those in excess of it. When used as a vacuum gauge for the condenser of a steam-engine, the tube becomes more flattened and curves inwards when the interior is exhausted. The action is the converse of that described, but the principle is just the same.

THE LAW OF BOYLE.

128. WE pass on to examine the two fundamental laws which govern the pressure of gases : the first is that of Boyle, or Mariotte ; the second is that of Charles, or Gay Lussac.

1. Boyle's Law. *The pressure of a portion of gas at a given temperature varies inversely as the space it occupies.*

Take a long glass siphon tube A C B, with parallel legs open at B, and either sealed or fitted with a stopcock at A Having placed it with the axis of either tube vertical, pour a small quantity of mercury into B C, and by opening the stopcock or withdrawing some air from A C, make the mercury stand at the same level E D in both branches.

The air in A D being now completely separated from that in B C, pour in more mercury *slowly*, and the level in the two legs will take the positions Q, and P R, where P R is a horizontal plane touching the surface in A C. Let K be the area of an internal section of the tube, then

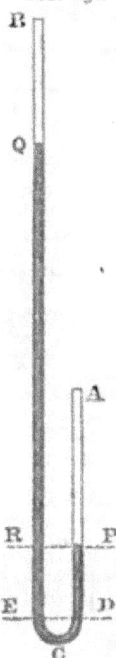

FIG. 138.

$$(\text{press. at } P)\, K = (\text{press. at } R)\, K$$
$$= \text{weight of } R\,Q + (\text{press. of atmosph.})\, K$$

Let h be the altitude of the mercury in a barometer at the time of making the experiment,

then $\dfrac{\text{pressure at } P}{\text{pressure of atmosphere}} = \dfrac{h + R\,Q}{h}$.

Now the air in A D has been compressed into the space A P by the additional weight of Q R, and it is found, by carefully weighing the quantities of mercury which the portions A D and A P would contain, that

$$\frac{\text{volume } A\,D}{\text{volume } A\,P} = \frac{h + R\,Q}{h}.$$

Hence $\dfrac{\text{volume } A\,D}{\text{volume } A\,P} = \dfrac{\text{pressure of air in } A\,P}{\text{pressure of atmosphere}}$.

In like manner, if the tube C A were made equal to C B and we began by filling both portions with mercury nearly up to the level of B, we should commence, as before, with air under atmospheric pressure ; some mercury might then be withdrawn from C B, the air in A P would become rarefied, and the law could be verified for pressures less than that of the atmosphere, just as in the first case.

Note. Since the *pressure* of a portion of gas varies *inversely as its volume*, and since the *density* of the same portion also *varies inversely as its volume*, it follows that *the pressure of a portion of gas varies directly as its density.*

Hence the law of Boyle is expressed by the formula

$$p = \mu\rho,$$

where p is the pressure of the gas, ρ its density, and μ a constant to be determined by experiment.

It is stated by Mr. Maxwell that the law under discussion ' is not perfectly fulfilled by any actual gas. It is very nearly fulfilled by those gases which we are not able to condense into liquids,' and further, that when a gas is about to pass by condensation into a liquid form ' the density increases more rapidly than the pressure.' For all practical purposes, where an engineer has to employ compressed air, the law may be taken to apply strictly.

Ex. 1. A vessel contains a quantity of air which weighs 8 grains, and exerts a pressure of 16¼ lbs. per square inch. If 3 grains more of air at the same temperature and pressure are introduced into the vessel, what pressure will now be exerted on its sides? (Science Exam. 1873.)

Ex. 2. The area of the cistern of a barometer is 4 times that of the tube, the mercury stands at 30 inches, and the whole length of the tube above the mercury in the cistern is 32 inches. A mass of air is now introduced, which would fill one inch of the tube at atmospheric pressure. Find the air space at the top of the tube.

This question takes in the correction for the alteration of level of the mercury in the cistern, which has to be made for an ordinary barometer as well as for the barometer gauge.

Referring to Fig. 132, we have A B = 32, A C = 2.

After the air is introduced, let C'D' be the length of column supported, also let A C' = x ;

Then $\dfrac{\text{pressure of air in A C'}}{\text{atmospheric pressure}} = \dfrac{1}{x}$, (by Boyle's law).

Therefore the pressure of the air in A C' supports $\dfrac{30}{x}$ inches of mercury.

Also the mercury sinks $x - 2$ inches in the tube, and therefore it rises $\dfrac{x-2}{4}$ inches in the basin. $\therefore \dfrac{30}{x} + 32 - x - \dfrac{x-2}{4} = 30.$

A quadratic equation, one root of which is $x = 6$, and the mercury falls through 4 inches.

Ex. 3. A cylinder, 20 feet long, is half filled with water, and inverted with the open end just dipping into a vessel of water. Find the altitude of the water in the cylinder ($h = 33$). *Ans.* 7·25 feet.

129. The diving bell is a loaded chest, the weight of which is greater than that of the water it would contain, suspended by a chain with the open mouth downwards. When the bell is lowered into the water, the air within it becomes somewhat compressed, (a fact which is easily verified by dipping an inverted empty glass tumbler into water,) and an air space is preserved which enables workmen to carry on their operations. The bell is supplied with fresh air by a pipe connected with an air pump, and may be entirely emptied of water by the air forced in by the pump.

The force tending to lift the bell is the weight of the water which the enclosed air displaces. Hence the tension on the suspending chain would increase as the bell descended in virtue of the diminution of air space due to the increased pressure. In estimating the buoyancy of the apparatus it is quite unnecessary to regard the pressure of the air within the bell, except so far as that pressure reveals the volume by the enclosed air.

The connection between the lifting force on the bell and the volume of the water displaced by the enclosed air may be rendered very apparent by an experiment. A small glass globe is partly filled with water and immersed with its neck downwards in a tall jar filled nearly to the brim. The globe just floats at the surface of the water. A sheet of india-rubber is tied over the mouth of the jar and pressed down by the hand. The globe immediately sinks, and may be held in any position near the top or bottom by regulating the pressure on the air at the top of the jar. The increase of air pressure is felt throughout the whole mass of water within the jar, the result being that the bubble of air enclosed in the globe diminishes and that the buoyancy also becomes less. The globe

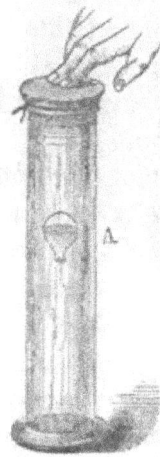

FIG. 139.

sinks, because less water is displaced, and it will rise again the moment the pressure is relieved.

130. *Prop.* To find the space occupied by the air in the bell at any depth below the surface.

Let A B C D be the bell, and draw E Q vertical, meeting the

FIG. 140.

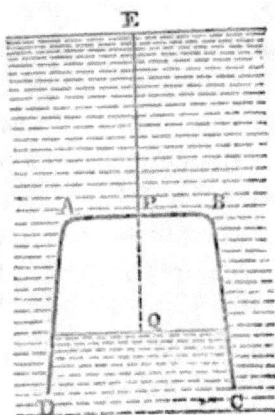

surface of the water outside the bell at E, and the surface inside the bell at Q. Let h be the altitude of the column of water in a barometer.

Then, according to Boyle's law, the pressure of the air within the bell is greater than that of the atmosphere in the proportion of volume A B C D : volume A Q B. But press. at Q : press. of atmosphere

$$= h + \text{E Q} : h.$$

$$\therefore \quad \frac{\text{volume A B C D}}{\text{volume A Q B}} = \frac{h + \text{E Q}}{h},$$

whence the volume A Q B is determined.

131. If more air be pumped into the bell, the pressure of the enclosed portion will be increased, and the water may be entirely forced out of the bell. Conceive now that a hollow cylinder is constructed of such a length as to reach to the foundations for the pier of a bridge, while its upper end remains above the surface of the water. If the cylinder be closed in at the top it will form an elongated diving bell, into which workmen can enter from above ; and it is clear that the principle under discussion may often be applied in this modified form with great advantage.

The use of a cylinder filled with compressed air was suggested by Lord Cochrane for working in wet ground and was first applied successfully in laying the foundations of the bridge at Rochester by Mr. Hughes. Mr. Brunel also employed cylinders from which the water was excluded by compressed air in forming the piers for the Saltash Bridge.

Here the operation **was** performed **on a very** large scale. A cylinder 37 ft. in diameter, and 90 ft. **in length**, was constructed of wrought iron plates, and lowered **through** 13 feet of mud till it rested with its base on **the solid rock.** Two engines of 10 horse-power were employed **to work the** air-compressing pumps. The cylinder, with **the** machinery inside, weighed 290 tons, but an additional load of 750 tons was required in order to keep it from rising. This shows the buoyancy due to displaced water. From 30 to 40 men could work inside the cylinder, and the pressure of the air to which they were subjected was equivalent to about 86 feet of water when the tide was at the highest.

The principle of the diving bell is also applied in diving dresses. The diver is clothed in a watertight dress fitted with a helmet and is supplied with air by means of a pump. **There is** an escape valve also, whereby the circulation of **fresh** air is maintained. The diver may be weighted up to 200lbs., but on closing the escape valve, he can rise at once **to** the surface in virtue of the buoyancy due to the increased displacement of water by the enclosed air.

Ex. 1. The weight of a diving-bell is 10 cwt., and the weight of the water it would contain is 6 cwt. Find the tension of the rope when the level of the water inside the bell is 17 feet below the surface, (h = 33 ft).

Here the air in the bell supports 33 + 17 feet of water.

$\therefore$ weight of water displaced by the air in the bell : 6 = 33 : 33 + 17,

$$\text{whence tension of rope} = 6 \cdot 04 \text{ cwt.}$$

Ex. 2. A cylindrical diving-bell of height (a) is sunk in water till it becomes half full. Show that the depth from the surface of the water to the top of the bell is $h - \dfrac{a}{2}$.

Ex. 3. A cylindrical diving-bell, of which the height inside is 8 ft., is sunk till its top is 70 feet below the surface of the water. Find the depth of the air space inside the bell (h = 33 feet).

Let x be this depth, then $\dfrac{70 + 33 + x}{33} = \dfrac{8}{x}$, $\therefore x = \dfrac{5}{2}$.

132. Boyle's law has been applied in the construction of an apparatus for ascertaining the depth of soundings at sea, without having any regard to the length of line paid out. The sinker is a hollow vessel with a strong glass window, and a small air pipe leads from the bottom to very nearly the top of the inner chamber. Since the compression below the surface of the sea increases with the depth, it is clear that the amount of compression of the air within the vessel will indicate the exact depth to which it has been sunk. The liquid pressure begins to act on the air in the tube, and soon compresses it sufficiently to allow some water to enter the chamber. The amount which enters will indicate the diminution of volume of the enclosed air, due to pressure, from which the depth can be as ertained. Since the air tube reaches to the top of the chamber, no water can escape while the instrument is being drawn up, and the reading of a scale seen through the glass window shows the depth of the sounding in fathoms.

THE MANOMETER.

133. This air-vessel, which registers the depth by applying Boyle's law, is merely one form of a *manometer* or apparatus for registering pressures of gases or liquids by observing the

FIG. 141.

amount of compression of a mass of air enclosed in a tube. The instrument shown in the sketch may be taken as an example. It consists of a barometer tube dipping into some mercury contained in a covered vessel communicating at D with fluid under pressure. When the pressure inside the closed vessel is equal to that of the atmosphere, the mercury in the tube drops to the level of that in the basin ; whereas the column of mercury rises as the pressure at B rises, and the air in the tube becomes denser according to Boyle's law. The exact pressure of the fluid can be determined by remembering that the pressure of the air on a section of the tube A C + the weight of the sustained column of mercury is equal to the fluid pressure on the same section of the tube.

Ex. Let a be the length of the column of air in the tube at the atmospheric pressure, and let A C become x when the air is compressed by a pressure p. Again let K, k be the effective areas of the basin and tube.

When the mercury rises through $a - x$ in the tube, it falls through a depth y in the basin, such that $k(a - x) = Ky$.

Let h be the altitude of the mercury in a barometer, then

$$\frac{p}{\text{press. of atmosphere}} = \frac{a}{x} + \frac{a - x + y}{h},$$

$$= \frac{a}{x} + \frac{a - x}{h}\left(1 + \frac{k}{K}\right).$$

If the ratio of p to the atmospheric pressure be assigned beforehand, we can solve this quadratic equation and ascertain x, and thus the tube may be graduated.

THE LAW OF BOYLE EXHIBITED BY A CURVE.

134. Take O X, O Y rectangular axes, call O X the line of volumes, O Y the line of pressures, and conceive that a mass of air occupying a volume v in a cylindrical vessel, fitted with a piston, exerts a pressure p on the sides of the vessel.

FIG. 142.

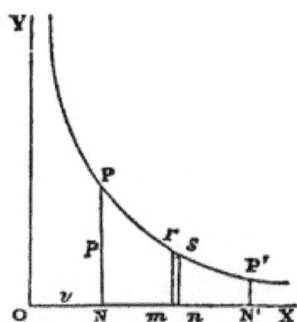

Let the piston move so as to expand the air to a volume v' under a pressure p'. It is evident that v and v' are proportional to two lines, viz., the distances from the base of the cylinder to the piston in the two positions, and that p and p' are also represented by straight lines. Hence we may take O N $= v$, P N $= p$, and assume that v, p are the rectangular co-ordinates of a point P referred to axes O X, O Y.

Similarly, v', p' are the co-ordinates of a point P', and the curve P P' represents the relative changes of volume and pressure.   By Boyle's law we have $pv = $ a constant quantity, and the curve possessing the property that O N × N P is constant at every point if it is known to be an hyperbola, or the section of a right cone, of proper shape to have the asymptotes at right angles, made by a plane parallel to its axis.   *Hence the curve representing Boyle's law is an hyperbola.*

DIAGRAM OF WORK DONE IN COMPRESSING A GAS.

135. Conceive that the gas is compressed from a volume v' and pressure p', to another volume v and pressure p, and that it is required to represent the work done during this compression by a diagram.*   Take r, s, two points in the curve P P' very close together, draw $r m$, $s n$ perpendicular to O x, and suppose that the pressure of the gas does not sensibly vary during its compression from a volume O n to a volume O m.

Since the space through which the piston moves is proportional to the difference of volume, it appears that the rectangle $s m$ represents the product of the pressure on the piston into the space through which it is moved during the compression from a volume O n to a volume O m, that is, the rectangle $s m$ represents the work done; and the same is true at every other point; hence the area P N N' P' represents the work done in compressing the gas from a volume v' and pressure p' to another volume v and pressure p.

ON THE MEANING OF THE TERM ELASTICITY.

136. Gases have often been distinguished from liquids by their behaviour under compression, and Professor W. H. Miller defines *elastic fluids* as those the dimensions of which are increased or diminished when the pressure upon them is diminished or increased.

We may now endeavour to obtain a definite conception of

---

* *The temperature of the gas is supposed not to change during the compression.*

elasticity as a property of matter, whether in a solid, liquid, or gaseous state. It may be shown by experiment that all substances are compressible to some extent, and we conclude that the minute molecules which form any substance, such as a ball of glass, are not in absolute contact, but are separated by certain intervals or intermolecular distances, which increase or diminish under the action of external force. It is further evident that the particles of the glass resist in a very high degree the action of any forces which tend either to separate them or to bring them closer together. They yield, no doubt, to every pressure, but the movement is extremely minute. When relieved they return at once to their normal positions, and they exert a strong effort of restitution, in virtue of a property which is termed *elasticity*.

We are now dealing with gases, and may define their elasticity as follows :—

Def. *The elasticity of a gas* **under any given conditions is the** *ratio of any small increase of* **pressure to the cubical com-***pression thereby produced.*

The term *cubical compression* denotes the ratio of the diminution of volume to the original volume, and Mr. Maxwell shows that the elasticity of a perfect gas is numerically equal to the pressure when the temperature remains constant. It is the practice, therefore, to use the term *elastic force*, meaning thereby *pressure*, as identical with the elasticity of a gas. We shall presently show that *every gas has two elasticities*, one *real*, the other *apparent*.

THE LAW OF CHARLES, OR GAY-LUSSAC.

137. We have next to examine the action of heat upon gases, and to point out that, whatever theory may be adopted to account for gaseous pressure, the fact of its dependence on temperature is thoroughly established.

Let the student hold a shrivelled bladder containing air before the fire ; the bladder will swell and become tense. Here the enclosed air expands under a constant pressure, viz., that of the atmosphere. If more heat be applied, the bladder may burst, the pressure still rising, although the volume of the enclosed air gets no larger. There are fundamental facts, which are now to be connected with the second law of gases.

2. The Law of Charles.— *When a portion of gas under a constant pressure is raised from the freezing to the boiling temperature, its volume will increase, by equal fractions of itself, for each degree of temperature; and this law holds, whatever be the nature of the gas.*

Conceive that a mass of air at the atmospheric pressure is enclosed in a tube of uniform bore by means of a drop of mercury, and that it occupies a length of thirty inches from the sealed end when cooled down to 0° C. Heat the tube to 100° C, and the air-space will elongate from 30 inches to 41 inches. The expansion is a little more than $\frac{1}{3}$ of the volume. The exact fraction was assigned by Gay-Lussac as ·375, but was corrected by Regnault, and is now taken to be ·3665.

Let v_0, v_t, be the volumes of a portion of a gas at temperatures o, t, respectively, and let $a = ·003665$, then the increase of volume from $o°$ to $t°$ *under a constant pressure* is $v_0\, a\, t$, therefore

$$v_t = v_0\,(1 + a\,t).$$

If the graduations of the Fahrenheit thermometer are adopted, we note that 180° F. corresponds to 100° C, and therefore that the expansion for 1° F. is $\dfrac{·3665}{180}$ or $\dfrac{1}{492}$ of the volume at 32° F. The more accurate value of the denominator is 491·13.

Hence $v_t = v_{32} \left(1 + \dfrac{t - 32}{492} \right) = v_{32} \left(\dfrac{460 + t}{492} \right),$

where t is the temperature on Fahrenheit's thermometer.

Ex. 1. A mass of air at 50° F. is raised to 51° F., what is the increase of its volume under a constant pressure ?

Here $v_{50} = v_{32} \left(\dfrac{460 + 50}{492} \right) = v_{32} \times \dfrac{510}{492}.$

$$\therefore \frac{v_{50}}{510} = \frac{v_{32}}{492},$$

or the gas expands $\frac{1}{510}$th of its volume at $50°$ for a rise of $1°$ F.

Ex. 2. In the Haswell colliery, near Newcastle, the depth of the upcast shaft is 936 feet, and its mean temperature is maintained at $163°$ F. by means of a ventilating furnace, while the mean temperature of the downcast shaft is $50°$ F. What work is done by the furnace, which sends 94960 cubic feet of air per minute through the mine at a temperature of $50°$?

By the last example, a column of air 936 feet high and at a temperature of 163 F., in the upcast shaft would occupy a height x when cooled down to $50°$ such that $x \left(1 + \dfrac{113}{510} \right) = 936.$ Whence $x = 766$ feet nearly, and the furnace does the work of lifting the air which traverses the mine through the height of $(936 - 766)$, or 170 feet.

The quantity of air passing up the shaft in one minute is 94960 cubic feet at $50°$, which weighs about 7407 lbs., the weight of 1 cubic foot of air at $50°$ being taken as ·078 lbs. Therefore

the work done $= 7407 \times 170$ ft. lbs.,
$= 1259190$ ft. lbs.,
$= 38 \times 33000$ ft. lbs. nearly.

The amount of coal consumed per minute is 8 lbs., which gives a duty of 17,628,660 ft. lbs. per 112 lbs. of coal.

138. *Prop.* To find the general relation between the pressure, temperature, and density of a portion of gas.

Let p_t, t, ρ_t be the pressure, temperature, and density of a portion of gas.

Then ρ_t varies as p_t when t remains constant (Boyle's law).

Also when p_t remains constant, we have, by Charles' law,

volume of gas at $t° = (1 + a t) \times$ volume of gas at $0°$,

or density at $0° = (1 + a t) \times$ density at $t° = \rho_t (1 + a t)$.

$\therefore \rho_t$ varies as $\dfrac{1}{1 + a t}$ when the pressure is constant.

But it is a principle in algebra that if x varies as y when z is constant and x varies as z when y is constant, then x varies as $y z$, when both y and z vary. Therefore, when both the pressure and temperature vary,

$$\rho_t \text{ varies as } p_t \times \frac{1}{1 + a t},$$

$$\text{or } p_t \text{ varies as } \rho_t (1 + a t),$$

or $p_t = \mu \rho_t (1 + a t)$, where μ is a constant. . . . (1).

Cor. 1. If the volume, and therefore the density, remains constant, while the temperature rises, the pressure will also rise; thus $p_t = \mu \rho_t (1 + a t)$,

and $p_0 = \mu \rho_t$, when $t = o$, since ρ_t does not change.

$$\therefore p_t = p_0 (1 + a t) \ldots (2)$$

It should be carefully noted that when a mass of air whose temperature is $o° c$ is heated to $100° c$, its pressure is raised from 1 to $1·3665$, being a rise of $·003665$ for each degree of temperature.

Cor. 2. If Fahrenheit's scale be adopted, formulæ (1) and (2) become respectively,

$$p_t = \mu \rho_t \frac{460 + t}{492}, \quad p_t = p_0 \left(\frac{460 + t}{492} \right).$$

Ex. 1. A certain volume of dry air weighs 50 grs. when the temperature is $o° c$, and the pressure 30 inches of mercury. What would be the weight of an equal volume of air at $25° C$ under a pressure of 40 inches of mercury? (Science Exam., 1871.)

Ex. 2. A cubic foot of air at a temperature of $100° F.$, and under a pressure of $29\frac{1}{2}$ inches of mercury, is cooled down to $40° F.$ and compressed by an additional $10\frac{1}{2}$ inches of mercury : find the volume.
Ans. $1137·86$ cubic inches.

Ex. 3. If 200 cubic inches of air at $60° F.$ under a pressure of 30 inches of mercury be heated to $280° F.$ while the pressure is reduced to 20 inches, find the volume, v, under the altered conditions. Taking the formula $p_t = \mu \rho_t \left(\frac{460 + t}{492} \right)$ and remembering that the volume of a portion of air varies inversely as its density, we have

$$\frac{30}{20} = \frac{v}{200} \cdot \left(\frac{460 + 60}{460 + 280} \right),$$

$$\therefore v = \frac{3}{2} \times \frac{740 \times 200}{520} = \frac{11100}{26} = 426·9 \text{ cubic inches.}$$

There is one remaining subject of much practical importance, which we can only touch upon slightly, and that is the
behaviour of a gas when compressed *suddenly* instead of
slowly. We say '*suddenly*,' because, by a quick action,
a result may be produced which would otherwise escape
notice, and the temperature will no longer remain constant.
The reason for the precaution mentioned in *Art.* 128, where
the mercury was poured in *slowly*, will now be made apparent.

ON THE HEAT DEVELOPED BY THE SUDDEN COMPRESSION OF AIR.

139. It is an experimental fact that heat is developed by
the sudden compression of air, and that a stream of compressed air when issuing from a closed vessel is sensibly
chilled.

Expt. 1. Take a tube 6 inches long, closed at one end,
and having a piston attached to a rod somewhat longer than
the tube.

If a small piece of German tinder be attached to the
piston, and the air in the tube be suddenly compressed by
driving the piston forcibly down, the tinder will probably be
ignited. In the same way the vapour of bisulphide of carbon
may be set on fire by compression in a glass tube, and
a flash of light will be seen when the piston reaches the
bottom of the cylinder.

Expt. 2. If air be pumped into a closed vessel the vessel
itself will become heated. This is very apparent in an air-
gun, where the compression is carried to a considerable extent. After the vessel of compressed air has cooled down
to the normal temperature, open a stop-cock fitted to it,
and allow the air to escape against the face of a thermo-
electric pile connected with a galvanometer. The needle
will swing violently round in the direction indicating that
the face of the pile has been chilled.

These are lecture-table experiments, but the same facts

are observed when compressed air is employed to drive an engine.

The earliest application of the use of compressed air for driving machinery in mines was made at the Govan colliery, near Glasgow, in 1849. A steam-engine was employed to compress air to a pressure of 30 lbs., and the air was then conveyed down the shaft to a winding engine at a distance of about 700 yards. The first difficulty met with arose from the heating of the compressed air in the cylinders of the compressing pumps, whereby it became necessary to keep the exit-valves flooded with water. There was a second difficulty underground, as the chilling produced by the sudden expansion of the air sometimes caused so great a degree of cold that the engine was stopped by the formation of ice in the cylinder and exhaust-pipe. As regards the loss of work by a waste of heat, it is clear that all cooling of the compressed air is a direct loss of power. This is one of the inconveniences attending the conveyance of motive-power by means of air under pressure. Furthermore, the chilling which is observed in the underground engine is a direct consequence of the work done by expansion behind a working piston.

This subject may be further illustrated by means of a diagram. Let A X be the line of volumes, A Y the line of

FIG. 143.

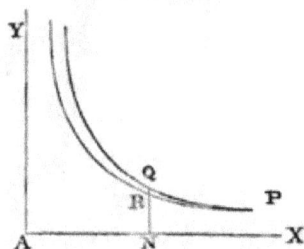

pressures, and take the case of a portion of gas whose volume and pressure are represented by the co-ordinates of P. Let P R represent the curve of pressures according to Boyle's law, then P Q will represent the same curve when no heat is allowed to escape. The gas is really more elastic than we should have supposed it to be. It has, in truth *two elasticities*, represented to the eye by the curves P R, and P Q. and we lose sight of the true elasticity, viz., that indicated by P Q, because it so rarely influences any observed result. This fact is illustrated in the propagation of sound, and a remarkable error was made by Newton in calculating the velocity of sound. The sound-waves compress and rarefy the air suddenly, and its elasticity has therefore the second, or true value. If this increased elasticity be disregarded, the velocity of sound in air, calculated in accordance with Boyle's law, will be less than the observed velocity by about one-sixth part.

THE ABSOLUTE ZERO OF TEMPERATURE.

140. Finally, we have to remark that in the applications of the modern theory of heat, the zero of temperature is taken to be the *real zero*, or that indication which an air-thermometer would give if the air were deprived of all its heat. There is no hope of ever depriving the air of all its elastic force, which would be the result of abstracting all its heat, but we nevertheless measure temperatures from their so-called *absolute zero*. In Mr. Maxwell's book the zero-point is shown to be $-273°$c, and the lowest observed temperature is stated to be $-140°$c. That is as far as anyone has yet gone in descending the scale of temperature. The student should also study that portion of the text-book on heat which explains the manner in which the laws of Boyle and Charles may be combined into one simple law, viz., *that the product of the volume and pressure of any gas is proportional to the absolute temperature.* As he may feel an interest in knowing exactly how much a gas is heated by compression, we give a formula in the first example which can be readily applied, and which shows the use that is made of the reading of the absolute zero.

Ex. I. A mass of air at 0° C is suddenly compressed till its pressure is increased tenfold: find the rise in temperature.

The formula referred to is $\frac{t'}{t} = \left(\frac{p'}{p}\right)^{.29}$, where p, t, are the pressure and temperature of the air at first, and p' t' the same quantities after the compression. Here $p' = 10\,p$, therefore

$$t' = t\,(1.953).$$

Now . represents 0° C, therefore t = 273° reckoning from the true zero, and consequently t' = (1·953) × 273° = 533°. This 533° is 260° above 0° C, and therefore the temperature of the air is raised 260° C or about 468° F. by the compression.

Ex. 2. The ventilation of the Haswell colliery involves an illustration of the use of this absolute zero. (*See* p. 121.) One part of the solution required us to ascertain the extent to which a column of air 936 feet high, at a temperature of 163° F., would contract in cooling down to 50° F.

Reckoning from the absolute zero, 163° F. is 460° + 163° or 623°; so also, 50° F. is 460° + 50°, or 510°. Now, the air contracts in proportion to the diminution of temperature from the true zero,

$$\text{therefore } \frac{623}{510} = \frac{936}{x}, \text{ and } x = 766, \text{ as before.}$$

CHAPTER VII.

ON PUMPS.

141. It is said that the suction-pump was invented 200 years before the Christian era, but its action was not understood until after the application of the barometer-tube for measuring atmospheric pressure. If the air be exhausted or sucked out in some manner from the top of a pipe dipping into water, the pressure of the atmosphere will force the water up the pipe, and will sustain a column of about thirty-two feet in height. If some of the water so raised be removed, more will be pressed up to supply its place, and in this way the action of a suction-pump is dependent on the pressure of the atmosphere.

Let A B, B C be two hollow cylinders, P a piston fitted with a valve opening upwards, and worked by a rod P D ; B a valve opening upwards, and let B C be less than the height of the column in a water-barometer, otherwise the water would never reach the valve B.

Suppose P to be at B, and the whole pump to be full of

air, the level of the water will be the same within and without
B C. Let the piston be raised, then the valve P will close,
and the valve B will open, because the

Fig. 144.

pressure of the air in B C is much greater than
that in P B. Hence the water will rise in B C
to some level *e*. On the descent of the
piston, P opens, B closes, and the air in P B
escapes through P, the column C *e* remaining
stationary. On the next ascent of the piston
the action before described is renewed, until
finally the water rises through B. It is then
compelled to pass through the valve P, and
is *lifted* by the piston till it escapes at the
spout E.

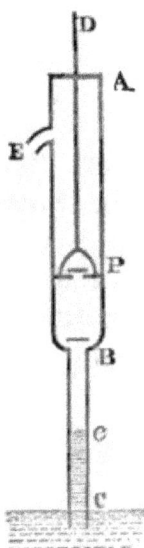

Note.—The pump-rod does the work of
lifting a column of water whose altitude is C *e*,
and base the sectional area of the cylinder A B.
The weight of this column is therefore the
tension of the pump-rod. This assertion may be questioned,
by reason that the weight of the column actually raised depends on the area of the tube B C, which is less than
that of A B. But the truth is that the shape or size of B C
has nothing to do with the tension of the pump-rod. If B C
were a cone with a wide mouth at C, the tension would be
just the same, and the pump-rod is always lifting a column
of water whose base is the area of the cylinder A B, and
height the altitude of the water lifted.

Prop. To find an expression for the tension of the rod in
a *lifting pump*.

Let C *e* $= 12x$ inches, and let $12h$ be the height of the water-
barometer, A the sectional area of A B in square inches, and
w the weight of a cubic inch of water in pounds, then

press: of air on upper surface of P $= 12h$ A w.

press : of air on lower surface of P $= 12h$ A $w - 12x$ A w.

∴ tension of pump rod $= 12x$ A w.

But $h : x = 15$ lbs. $: 12 x w$,

$\therefore 12 x w = \dfrac{15 x}{h}$, and tension of rod $= \dfrac{15 x A}{h}$ lbs.

This formula holds when the water has risen above P.

THE PLUNGER PUMP.

142. The plunger pump is the same apparatus as that already described, so far as raising the water into the cylinder is concerned, but it differs in the mode of lifting the water afterwards : the bucket is replaced by a solid plunger or piston P C, shown in the diagram. There are two valves, A and B, each opening upwards, and the water fills the cylinder as the plunger P C moves to the right, the valve B opening and A remaining closed. On the return of the plunger, B closes, A opens, and the water in F D is forced through the valve A. A pump so made will propel the water to a considerable height, and is used in fire-engines, the common name for it being a *force-pump*. Since the water is urged forward only at each return stroke of the plunger, the action is intermittent, but it may be made continuous by the use of an air-vessel, that is, a chamber enclosing air, into which the water is pumped. The air becomes compressed, and its elastic force drives the water through an exit-pipe in a continuous stream.

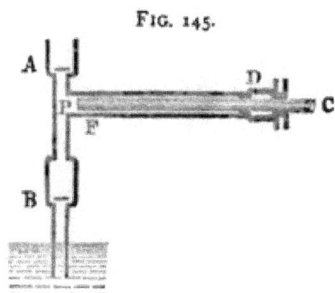

Fig. 145.

THE PLUNGER AND BUCKET PUMP COMBINED.

143. In order to obtain a continuous flow of water without the use of an air-vessel, the plunger and bucket pump may be combined in the same cylinder. The drawing shows the up stroke, when the bucket P is lifting the water, and compelling it to pass through the exit-pipe E. The valve B is open, because the water from the cistern is following the bucket during its ascent. On the descent of the pump rod the plunger Q will enter the cylinder D B, the valves at P will open, and the plunger will be forced down into a cylinder full of water, and closed at the bottom by the valve B. Hence the plunger will displace a quantity of water, which must pass into the exit-pipe E. The pumping action is therefore continuous.

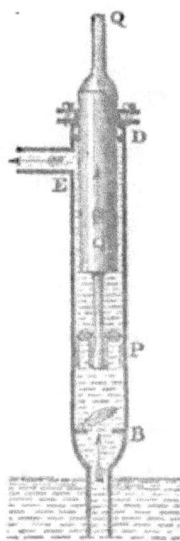

FIG. 146.

144. In pumping water from a mine, it is a common practice to combine on the same rod a plunger and a bucket pump at considerable intervals from each other. In the Clay Cross colliery, for example, the water is raised from a depth of more than 400 feet. The main pump rods, or *spears*, are made of pitch-pine, in lengths of 45 or 50 feet, and are 16 inches square. The bucket of the lifting pump raises the water into a drift about 150 feet above the bottom of the mine. A plunger is attached to the pump rod and works in a cylinder on one side of it, thereby forcing the water already lifted up the remaining 270 feet. There is also a second plunger attached to the pump rods, which forces up more water, coming from an independent supply at a depth of 217 feet. Each pump delivers somewhat more than 100 gallons of water per stroke. The object of combining the forcing and lifting pump is to economise work. The pump rods, with the iron-work attached, are of great weight, and they must be lifted through the stroke of 10 feet in order to raise the water at all.

It would be absurd to sacrifice the work done in lifting the weight of the pump rods, and accordingly they are armed with plungers, and made to do useful work by forcing up water during their descent.

PUMP USED IN WELL-BORING.

145. In boring holes for Artesian wells it often happens that after the water-bearing strata have been pierced the level of the water in the bore remains at some distance below the surface of the ground, and must be brought up by a pump. The bore is a few inches in diameter, and the last mentioned combination is inadmissible; nevertheless the discharge may be rendered continuous by proceeding on a different principle.

The pump is a cylinder, say 12 inches in diameter and 12 feet long. Two lifting buckets, A and B, are fitted in the cylinder, and each bucket has a valve opening upwards. These buckets are made to come together and separate, the upper one being attached to a hollow tube or p.pe, through which the solid rod attached to the lower bucket can move freely. When the buckets approach the valve A opens, and B closes, whereby B acts as a pump lifting the water through A. When they separate B opens and A closes, whereby the water passes through B and follows A, and thus the lower bucket acts as a pump or valve alternately. It may be arranged by suitable mechanism that one bucket shall stop a little before the other, and the result is that the column of water raised is never quite at rest, and that the buckets are relieved from the shock due to the inertia of a heavy mass of water.

THE AIR-PUMP.

146. The exhaustion of air from a closed vessel is the same mechanical operation as the pumping of water. The common pump does the work of exhausting air from B C until the water has risen above B. Hence an ordinary air-pump is the same apparatus as a suction-pump, the difference consisting mainly in the valves, and in the fact that it becomes necessary to counterbalance or remove the pressure of the air on the upper surface of the piston, otherwise the labour of working an air pump would be insupportable. It is an instructive experiment to ascertain by trial the force necessary to draw up a piston 6 inches in diameter from the bottom of an open cylinder into which it actually fits. Since the area is 28·27 square inches, the force required would exceed 415 pounds. In an air-pump the pressure of the air above the piston would be 14·7 lbs. on the inch, and that below the piston would soon fall to less than 1 lb., whereby the atmosphere would oppose the ascent of the piston with a pressure exceeding 14 lbs on the square inch, and would afterwards drive it down with the force of a blow.

It being now made clear that if a piston fitting in a cylinder be exposed on one side to the full pressure of the atmosphere and on the other side to a diminished air-pressure, it will be driven by a pressure which may be regulated by the exhaustion of the air, we may point out that the coining presses at the Mint are worked on this principle. The lever handle which rotates the screw of a press, and so impresses a steel die upon the blank piece of metal, must be pulled much more energetically for the coining of a sovereign than for that of a threepenny-piece. It is therefore connected with a piston dragged through a cylinder in one direction against the full pressure of the atmosphere, and then set free. The return of the piston is opposed by the diminished air-pressure, due to a partial vacuum kept up by an air-pump, and it follows that the force of the rebound will depend on the degree of exhaustion of the air, and may be adjusted with great nicety to the requirements of the press.

147. We proceed to examine certain forms of the air pump.

1. *Smeaton's air-pump.*

Here a hollow cylinder **A B** communicates with a receiver by a pipe c, and there are three valves, one at B, another in the piston P, and a third A at the top of the cylinder. When the piston moves from B to A, P closes, but A and B open, and

FIG. 147.

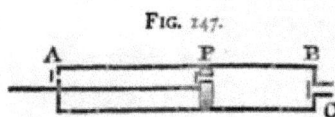

air from the receiver rushes into P B, while that in P A is expelled through A. When the piston returns A and B close, while P opens, and the air in P B is forced through the valve P, ready to be swept out on the next return of the piston.

Let A be the volume of the receiver, B the volume of the cylinder, ρ_1, ρ_2, the densities of the air in the receiver after 1, 2 strokes respectively, then

$$\rho_1 (A + B) = \rho A,$$
$$\rho_2 (A + B) = \rho_1 A, \text{ and so on.}$$
$$\therefore \rho_1 \rho_2 (A + B)^2 = \rho \rho_1 A^2, \quad \therefore \rho_2 (A + B)^2 = \rho A^2.$$

If this goes on for n strokes we have $\rho_n (A + B)^n = \rho A^n$, which gives the density or pressure of the air in the receiver after n strokes.

Note 1. The piston is relieved from the pressure of the atmosphere by the valve A.

Note 2. An air-pump valve is made of a small piece of oiled silk stretched over a hole in a plate and tied with a thread. The silk is lifted by the air sufficiently to permit its escape. When the silk is pressed on the plate the valve is perfectly closed.

2. *Hawksbee's air-pump.*

This is the form in common use; it consists of two cylinders, placed side by side, but open at the top to the external air. The air-pressure is nearly the same on both pistons, which are furnished with valves as in Smeaton's air-pump, and the piston rods are worked by a pinion placed between them and gearing with a rack on each piston rod. As one piston rises the other descends, the atmospheric pressure is counterbalanced, and the exhaustion is very rapid. But the degree of exhaustion is not great, for the valves in the pistons will soon fail to open, on account of the pressure of the external air overcoming that in the valve passages.

3. *Grove's air-pump.*

Here the valves B and P are abolished, and the valve A

only is left; the construction is more simple and the exhaustion is more perfect. The pipe
c is now brought to a point
intermediate between A and B,
whereby a solid piston does the
work of pumping. This artifice is to allow the piston to pass beyond c; the air in the receiver expands and fills the space between A and the piston; on the return stroke the enclosed portion of air in AP is swept out of the valve A. It is clear that the exhaustion will go on until the air compressed in the passage through the plate A becomes unable to lift the valve. The principle is an excellent one, and it only remains to improve on the valve A. This has been done in

Fig. 148.

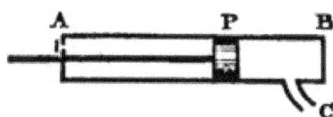

4. *Sprengel's air-pump.*

Here the cylinder A B is replaced by a glass tube, about 5 feet long, connected with a funnel A containing mercury. A piece of tubing at F is
tightened by a screw clip, so that the
flow of mercury down the tube can be
regulated. The tube A B dips into a
vessel of mercury, and is attached to
the receiver E by a pipe C. The mercury flows down in a series of drops,
P, Q, separated by intervals, and as soon
as Q has passed the neck c the air in
the receiver expands into the portion
C Q. This small portion of air is immediately cut off by the next drop P,
and can never return, and the receiver
is exhausted by continually wiping
out the portions of air which never
cease to enter the tube through the
neck c. This is Grove's pump under
an improved form, and the degree of
exhaustion which can be effected by it is truly marvellous.

Fig. 149.

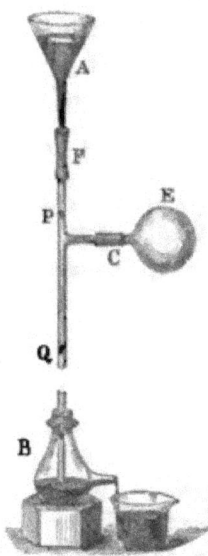

Note 1. In describing the air-pump we have said that the air rushes from the receiver into the barrel, and follows the piston during the stroke. A mass of air cannot move without the expenditure of heat, and the heat necessary for its motion is taken from the air itself. Hence the air in the receiver is chilled during the exhaustion, and a cloud of vapour is formed inside the receiver during the first few strokes. If the air be charged with moisture the effect is very striking.

Note 2. The pressure of the air in the receiver of an air-pump is measured by a gauge, that in common use being the barometer-gauge already described. If h be the altitude of the mercury in a barometer, z its altitude in the tube of the gauge, ρ the density of air, we have

$$\text{density of air in receiver} = \rho \, \frac{h - z}{h} \cdot$$

Another gauge is a fine syphon-tube A C B, closed at the end B, and connected at A with the receiver. The leg C B is about 4 inches long, and is filled up to B with mercury. As the exhaustion proceeds the pressure of the air in the receiver becomes unable to support the mercury in C B. The difference of level then shows the pressure of the air, and if the exhaustion were quite perfect the level of the mercury would be the same in A C and C B.

Ex. 1. An air-pump is so constructed that $\frac{1}{3}$ of the contents of the receiver is removed at each stroke. If the air before the first stroke is under a pressure of 30 inches of mercury, what is its pressure after the third stroke ?   (Science Exam. 1871.)

Ex. 2. If the volume of the receiver of an air-pump be 10 times that of the barrel, show that the density of the air in the receiver will be reduced one-half before the end of the eighth stroke.

Ex. 3. The mercury rises in the barometer-gauge of an air-pump through $6\frac{1}{2}$ inches in 8 strokes : compare the capacities of the receiver and barrel. ($h = 30$ inches.)   *Ans.* B : A = 31 : 1000.

Ex. 4. A thin bottle filled with air is placed in the receiver of an air-pump, and when the gauge stands at 21 inches the bottle bursts, whereby the mercury falls to 17 inches. Prove that the content of the receiver $= \frac{21}{4}$ content of the bottle.

THE CONDENSING SYRINGE AND BLOWING ENGINE.

148. The condensing syringe consists of a cylinder A B, open to the air at A, and having two valves, one in the piston opening towards B, and another at the bottom of the cylinder. As the piston moves from A to B the valve P closes, while B opens, and the air in P B is forced into a receiver con-nected with the pipe c. On the return stroke B closes, P opens, and the space B P again becomes filled with a supply of air, which is ready in its turn to be forced into the receiver.

FIG. 150.

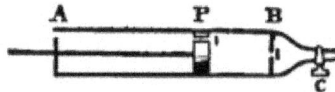

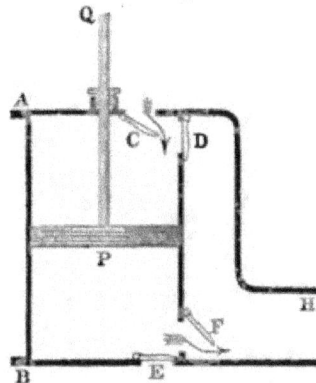

149. The supply of air to furnaces demands a special form of pump which is analogous to the pump for forcing water, and is a double-acting condensing syringe on a large scale. There are no valves in the piston, but there are four valves in the cylinder (as shown in the drawing), two opening inwards for the ad-mission of air, and two open-ing outwards for its exit. The pipe H leads to a receiver in which the compressed air is stored up, and it is clear that when the piston descends the valves c and F open, while D and E close, whereas on the ascent of the piston the reverse takes place. Thus, a cylinder full of air is swept into the pipe H at each stroke of the piston.

FIG. 151.

Some years ago a large blowing-engine was put up at the Dowlais Iron Works, in which the blowing cylinder was 12 feet in diameter, with a 12 feet stroke, and the piston, attached to the beam of an engine, made twenty double strokes per minute. The discharge-pipe H was 5 feet in diameter, and the air was compressed to a pressure of $3\frac{1}{4}$ lbs. on the square inch, the quantity discharged at this pressure being 44,000 cubic feet per minute. There is a model of this machinery in the museum of the School of Mines.

THE FORGE BELLOWS.

150. While describing these important machines, we must not omit a very homely example which illustrates a principle of action. The smith's bellows is shown in the sketch, and it is really the same apparatus as the accumulator to be described presently, the only essential difference being that it works with air instead of water. The ordinary bellows has

FIG. 152.

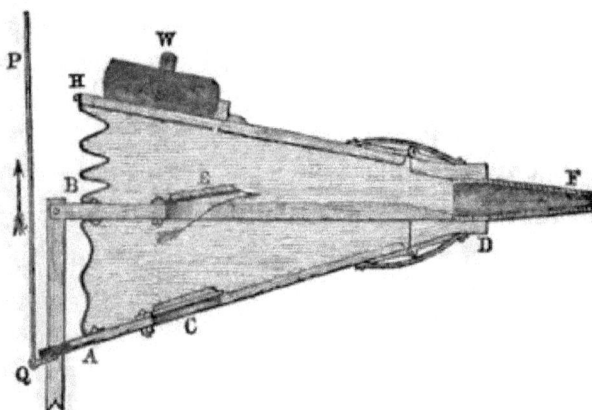

an intermittent action, and the object now is to make the blowing continuous. This is effected by a double chamber, the lower one corresponding to an ordinary bellows, and having a valve c opening upwards. When A D falls the valve c opens, and the chamber B A D becomes filled with air. When A D rises the valve c closes, E opens, and the air in B A D is forced into the chamber H B D, and escapes in a blast through the nozzle F. The weight w keeps the air in H B D under a pressure which may be adjusted, and the pipe F is contracted, so that the air escapes less rapidly than it is pumped in. Thus **the flow is continuous.** This is also the construction of an organ-bellows, where a continuous flow of air is required for sounding **the** tubes. The pressure of the air accumulated in the upper chamber can be adjusted by varying the load **w.**

CHAPTER VIII.

ON THE HYDRAULIC PRESS AND HYDRAULIC CRANES.

151. **The** hydrostatic **press is a** machine which, for power **and** simplicity of construction, is unequalled. In studying its action the student cannot fail to observe the wide difference between the unfruitful knowledge **of** a principle and **the** successful application of it in **practice.**

The principle of the press is obvious to anyone who understands the equal transmission of fluid pressure. Conceive that a closed vessel with its upper surface level is completely filled with water, and that two openings are made in it, which **are replaced by** pistons of areas 1 and **10** square inches. If a weight **of 1 lb.** be placed **on the** smaller piston, **a pressure** of 1 lb. **will be felt everywhere in** the interior of **the** fluid, and **the pressure on the larger piston** will be 10 lbs. Thus, a force **of 1 lb., acting on the area** 1 square inch, produces a pressure **of 10 lbs. on the area 10** square inches.

The same principle is **exhibited** in the *hydrostatic bellows*. That is the name given **to an** apparatus consisting of two boards connected like **a** bellows by water-tight flexible sides, **but** capable of being separated in parallel planes to a distance **of** 3 or 4 inches. A long slender **tube is fitted on** a pipe leading **to** the interior **of the** bellows. **A** heavy weight is **placed on** the boards, and **water is poured** into the tube. If the weight is not **disproportionate to** the size of the apparatus, it will **be found to rise when the** column of water in the tube gets to a moderate **height, and** in this way it is easy to arrange that a wine-glass of water shall support the weight of a man.   This is a lecture-table apparatus.   In order to obtain a working machine we must construct a cylinder of great strength, having a solid plunger, or *ram*, as it is called, passing into it through a water-tight collar, and we should rely upon an ordinary force-pump for putting a pressure upon the water in the cylinder.

152. The invention of the water-tight collar, which prevents any escape of water from the cylinder, was brought into use by Bramah.   No other arrangement of packing is worth anything under a water pressure, which often exceeds 2 tons on the square inch.   Two forms of packing are adopted in the press, (1) *the ring*, (2) *the cup-leather.*   A section of the ring is shown as enclosed in the chamber A B C D, D B being the side of the ram.   The water under pressure leaks into the chamber, and forces the leather tightly against the sides A C, D B.   The greater the pressure the more impossible it is for the water to pass beyond the ring.   The *cup-leather* is a simple cup fastened to one end of the piston, which acts

FIG. 153.

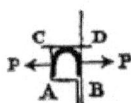

FIG. 154.

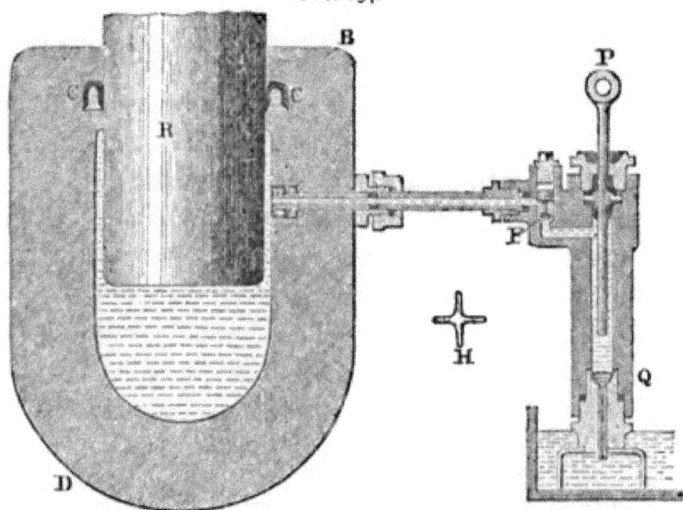

only when there is water pressure on the inside of the cup.

The press itself consists of a strong cast-iron vessel B D, with a nearly hemispherical base, into which a cylindrical ram R is inserted. The ram is encircled by a water-tight collar, and the work of the press is done by the ram. The hemispherical base is only important in very large presses, and the object is to obtain a sound casting without any line of division due to a want of uniformity in the flow of heat outwards. The large press for lifting the tube of the Menai Bridge bore a load of about 900 tons; it was cast with a rectangular base, and the bottom was forced out. It is a rule in mechanics that a vessel intended to support either internal or external fluid pressure should not be constructed with flat sides.

The remaining portion of the machine is a simple plunger pump P, having two valves, one at Q, the other at F, whereby the water is sucked up into the pump-barrel and then forced into the vessel D B. H is an enlarged section of that part of the valve Q, below the seat, which acts as a guide.

Let r = radius of ram, s = radius of plunger, P the pressure applied to plunger, W the resistance overcome by the ram. Then

$$\text{P} : \text{W} = \pi s^2 : \pi r^2 = s^2 : r^2, \therefore \text{W} = \text{P} \frac{r^2}{s^2}.$$

The plunger would in many cases be worked by a lever. Suppose that a force Q acting at an arm a of a lever produces a pressure P on the plunger, which is connected with the lever at a distance b from the fulcrum, then

$$\text{P}\,b = \text{Q}\,a, \qquad \therefore \text{W} = \frac{\text{Q}\,a}{b} \cdot \frac{r^2}{s^2}.$$

Note 1. As might be anticipated, the loss of power by the friction of the leather collars is not very serious. Mr. Hick, of Bolton, has made some experiments on this subject, and states that for rams of 4 inches diameter the friction is about $1\frac{1}{2}$ per cent. of the total resistance W, and for rams of 8 inches diameter it is about $\frac{1}{2}$ per cent., the water pressure being from 5000 to 6000 lbs. per square inch. Mr. Rankine, on the other hand, estimates the friction at $\frac{1}{10}$th the resistance.

Note 2. The principle of work maintains here as in all other machines. Let the plunger move through a space x while the ram advances through y ; since the volume of the water in the whole apparatus remains constant, we have

$$\pi s^2 x = \pi r^2 y, \qquad \therefore \quad r^2 : s^2 = x : y,$$
$$\therefore \quad \mathrm{w}\, y = \mathrm{p}\, x.$$

Ex. 1. In a press the plunger is $\frac{7}{8}$ inch, and the ram $9\frac{1}{4}$ inch in diameter. The plunger is worked by a lever ; the distance from the pump to the fulcrum is $3\frac{1}{4}$ inches, and that from the fulcrum to the power p is 78 inches : prove that w = 2682 p.

Ex. 2. The plunger is $\frac{3}{4}$ inch, and the ram 10 inches in diameter ; the arms of the lever are 6 feet and 1 foot. A weight of 20 lbs. is hung at the end of the lever : find the pressure on the ram.

Ans. 9 tons, 10 cwt., 1 qr., 25·3 lbs.

The hydrostatic press is employed by Sir Joseph Whitworth for compressing *ductile steel* while in a fluid state, and the pressure put on the metal amounts, in some cases, to 20 tons on the square inch. The principle of the press admits of extension as far as the strength of the containing sides of the pressing chamber will permit. Conceive that the ram of a press is 100 sq. inches in area, and that the pressure of the water is 2 tons per sq. inch ; the ram may be prolonged and reduced to 10 sq. inches in area, it would then exert a pressure of 20 tons per inch on the fluid steel contained in a compressing chamber. Lead pipes for conveying water, and the lead wire used for making rifle bullets, are examples of the flow of soft metal under hydrostatic pressure. The lead is allowed to solidify partially in a massive cylinder, and is forced out by pressure through an opening of the required form and made of the hardest steel. This is an instance of the flow of solids under pressure, all trace of fluidity having disappeared by the time the lead leaves the chamber. We pass on to enquire into the extended use which is now made of the press by working it with a diminished water pressure.

THE ACCUMULATOR FOR STORING UP THE PRESSURE OF WATER.

153. Many years ago, Sir W. Armstrong noticed a small stream of water which flowed from a great height down a steep declivity and turned a single overshot wheel at the bottom. Not more than $\frac{1}{20}$th part of the energy stored up in the water was utilised, whereas if the water had been enclosed in a pipe, and caused to act upon a suitable water engine, a large proportion of its power might have been usefully employed. This reflection led to the invention of a water-pressure engine, and ultimately to the application of water-power for working cranes.

In 1846 the first hydraulic crane was set up on the quay at Newcastle, and the pressure of the water was that due to a head of about 200 feet.

The new system was then tried at Great Grimsby on level ground, and, for some unexplained reason, a water-tank was placed on the top of a high tower in order to obtain the necessary pressure. But it soon became apparent that if water be pumped into a vessel fitted with a moveable plunger, and if the plunger be loaded with a sufficient weight, the enclosed water will be in just the same mechanical condition as if it were in communication with a reservoir some hundreds of feet above it. The water is under pressure in both cases, and it makes no manner of difference whether the pressure is caused by a load of iron or a load of water. With a properly constructed loaded vessel we can readily obtain a supply of water under a pressure of 700 lbs. on the inch, which corresponds to a natural head of about 1500 feet. The vessel in which the water is stored and confined has been called an *accumulator*, because it accumulates the power of a steam-engine, and becomes a storehouse of the work done by the steam.

The accumulator, *which is to be regarded simply as a weighted hydraulic press*, is a long iron pipe or cylinder A A, fitted with a solid plunger P P. The bottom of the cylinder communicates with a pumping engine on the one side, and with the cranes on the other side, as indicated by the pipes F, E, in the drawing. Water is pumped into the cylinder before the cranes are set to work, and is continually entering at one side F, and passing out at the other side E. The

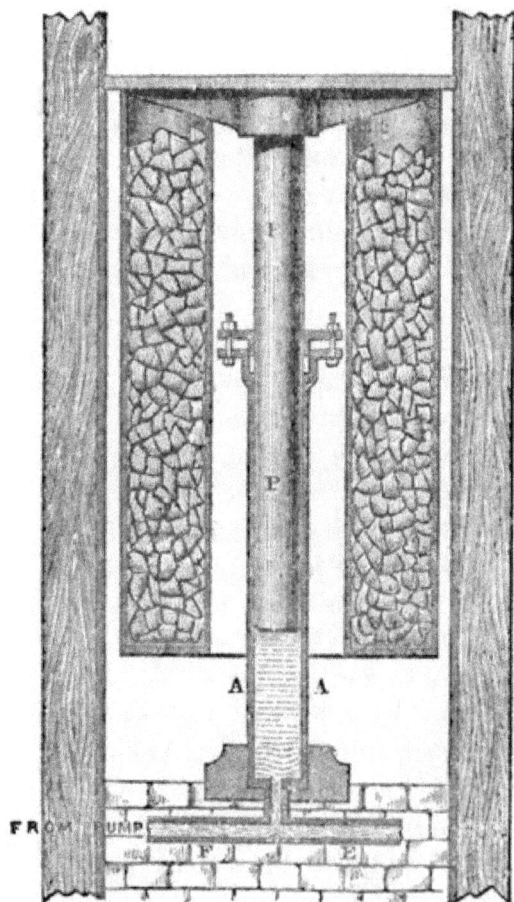

FIG. 155.

plunger passes through a water-tight packing or gland, and supports a hollow annular cylinder, which is loaded with scrap-iron and heavy waste metal. So long as the plunger is raised the whole of the water, extending to the most distant crane, is under the pressure determined by the area of P P and the load upon it. For example, if the diameter of the plunger be 17 inches its area will be 226·98, or 227 sq. inches ; and further, if the load be 70 tons, or 156,800 lbs., it is clear that the pressure per square inch on the water enclosed in the accumulator will be 156,800 ÷ 227, or 700 lbs., nearly. This is equivalent to taking the water from the bottom of a reservoir 1600 feet deep. It is usual to preserve this working pressure of 700 lbs. on the inch in all applications of water-power to cranes or analogous machinery.

In illustration of the practical value of hydraulic power, we refer to the corn warehouses at the Birkenhead docks, where the water under pressure is conveyed to various points through a space of 2300 feet. There are three accumulators, worked by an engine of 370 horse-power. The labour performed is that of hoisting grain out of vessels, and distributing it through the warehouses, the bulk to be dealt with being 250,000 tons per annum.

We must now endeavour to give some idea of the use of the accumulator, and shall describe a crane worked by water pressure. Other cranes have three oscillating cylinders, and are adapted for lifting weights up to 50 tons.

THE HYDRAULIC CRANE.

154 This crane is of the ordinary construction so far as the jib B L and tie rod A B are concerned, but there is no train of wheels, and the lifting chain passes through the crane post to the hydraulic cylinders. In the drawing we have shown a chain, fastened at one end to the solid abutment F, passing once round each pulley C and M, but running over two sheaves in succession at D. The pulleys at D are placed at the end of the ram of a press, and it is clear that when D moves through 1 foot the weight suspended at W will rise 4 feet. In this respect Sir W. Armstrong has inverted the principle of reduplication, and has taken the second system of pulleys in a reverse form, the larger power being employed to lift the smaller weight. The object is to gain in speed, and accordingly a heavy mass which six men could barely lift in a quarter of an hour may be run up by the crane in two minutes. We shall presently show that a

FIG. 156.

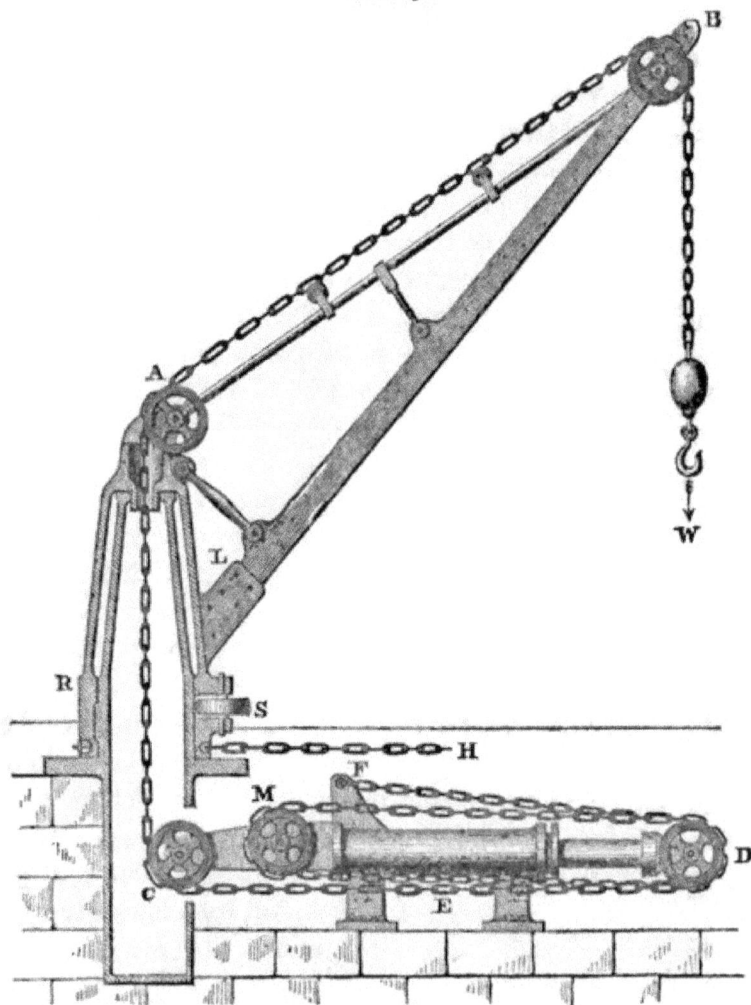

single cylinder may put forth two degrees of force, one greater than the other, whereby the power is economised when a small load is being lifted. In a 10-ton crane at Woolwich there are three cylinders, lying side by side, and capable of exerting three degrees of lifting power. The water may be admitted into the middle cylinder, into the two extreme cylinders, or into all three at once, and the powers are as the numbers 1, 2, 3. The lifting chain passes over three pulleys attached to a frame carrying the plungers, whereby the speed of the lift is multiplied 6 times. The diameter of each ram is $11\frac{5}{8}$ inch, the pressure on it being $700 \times 106\cdot139$ lbs., or about 33 tons, making 100 tons for the whole power which can be exerted. This has to be divided at once by 6, on account of the loss due to the pulleys. The travel of each ram is 9 feet, and the load can therefore be raised through more than 50 feet. The crane is worked by one man, who lets the water into the cylinders, and allows it to escape by merely pulling over a lever.

The rams carrying the pulleys and framework at D are caused to return by the pressure of the water on a piston 4 inches in diameter, which is not shown in the drawing. The force here exerted is $12\cdot566 \times 700$ lbs., or nearly 4 tons.

FIG. 157.

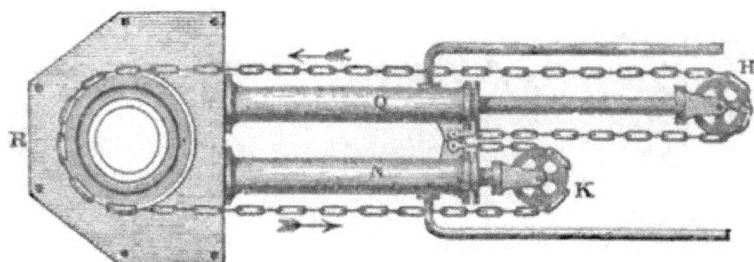

The whole crane rotates round the post, or is *slewed*, as it is termed, by the action of a separate set of hydraulic rams. The chain passes round the shell of the crane at R, and over two pulleys at K, H, the ends being attached to the framework. When the water is let into N and out of Q the pulley K will advance, and H will run back, and thus the chain will carry the crane round. A reverse action may be set up in like manner. The diameter of the rams for slewing are considerably less than those used for the hoist, and there is a similar sacrifice of power to a gain in speed. The slewing is effected by pulling over a hand lever. The pillar of wrought iron is some 18 to 20 feet high, and the radius of the sweep of the crane is 32 feet. There are friction rollers, whereof one is shown at s in the first diagram, which assist the slewing action.

Note. The principle of the crane may be applied under many forms ; thus a low-pressure hydraulic hoist has been put up at the Westminster Palace Hotel, the pressure being 35 to 40 lbs. on the square inch. The piston of the hoist is 20 inches in diameter, with 10 feet stroke, the height of the lift being 56 feet. The piston drives a rack, and thereby rotates a pinion placed upon the shaft of a large pulley which multiples the stroke of the piston $5\frac{1}{2}$ times, and causes the lift to rise or fall as required.

THE PRINCIPLE OF COUNTERBALANCING FLUID PRESSURE.

155. In applying water-power great use is made of the opposition of liquid pressure, and it may be an advantage to trace this idea in its application to purposes which are entirely different.

1. *By opposing fluid pressure we obtain two different powers in an hydraulic cylinder.*

Let A B represent an hydraulic cylinder, and conceive that the work is done while the piston P is being driven from

FIG. 158.

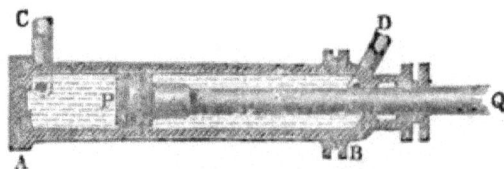

A to B. If the full pressure of the water be exerted on the side A P, while that on the side B P is removed, the engine will give out its greatest power. Whereas if the pressure on the side B P be opposed to that on the side A P the power exerted will be greatly diminished.

Let 2*a* be the area of the piston, and *a* that of the piston rod, *p* the pressure of the water, c and D two pipes, which may lead either to the accumulator or to the exhaust at the control of suitable valves.

If D and c be both open to the accumulator the pressure on the piston is 2 *p a* — *p a*, or *p a*. A water pressure pump works in this way, D being always open to pressure, and c opening to the exhaust and pressure alternately, whereby the pressure in each direction is *p a*. With c open to pressure and D to exhaust, the power would be doubled.

2. *By opposing fluid pressure an ordinary* **plunger pump** *may be made double instead of single acting.*

We have already seen that a lifting and plunger pump combined is double-acting, or forces the water at each stroke, but this compound pump does not work well at extreme pressures. In forcing water into the accumulator against a pressure of 700 lbs. on the inch an air-vessel is inadmissible, and in truth every precaution is taken to prevent air from entering with the water. It is therefore an advantage to convert the common force-pump into an apparatus propelling the water at each stroke. This may be done by opening a passage into the cylinder at the back of the piston, for it will be seen, on inspecting the diagram, that the piston P and the valves E, F, form an ordinary force-pump, and that the peculiarity in the arrangement is the open pipe c leading into one end of the cylinder.

Here A B is the pumping cylinder, P the piston, P Q the piston rod, whose sectional area is half that of the piston ; H a pipe leading to the accumulator, E and F two valves outside the pump, K the pipe leading to the reservoir from which water is drawn. As the piston moves from B to A the valve E opens, F closes, and the water in A B is forced into the accumulator ; that is, a quantity of water whose volume is half the content of the pump passes through H.

At the same time P B is filled with water at the ordinary pressure.

FIG. 159.

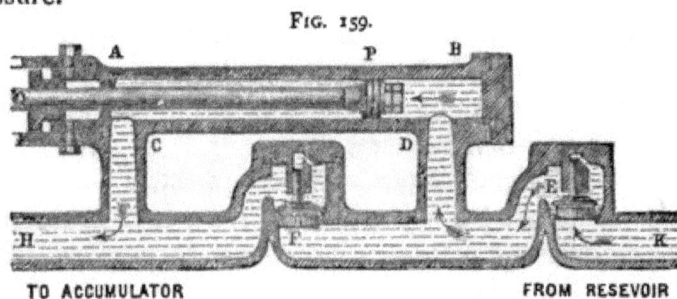

TO ACCUMULATOR   FROM RESEVOIR

When the piston returns from A to B the valve E closes, the water in P B is compressed; and as there is no escape except through F, its pressure immediately rises sufficiently to open F, and the water in P B is forced into the accumulator. But the space A P fills with water under pressure during this movement, and one-half the pressure on the piston P is counterbalanced. Hence one-half of the water contained in the cylinder is forced through H during the return stroke.

The doctrine of work is here admirably exemplified. It might be imagined that work was lost by the opposition of pressures; but no such loss really occurs, for the measure of the work done is the quantity of water which finally leaves the pump. If positive work be done on one side of the piston, negative work is done on the other side, and the effective sum is the difference of the amounts of work performed by the opposing forces.

3. *By opposing pressures we obtain a balanced valve which can be opened by a small force.*

The principle under discussion applies in every case of fluid pressure, whether the fluid be a gas or a liquid, and we may instance a disc valve which opens against the pressure of steam. With a single valve the whole pressure on its area must be overcome before the valve can be raised, whereas if two equal valves are threaded on the same spindle, the pressure on one disc counterbalances that on the other.

The valve, called a *double-beat* disc valve, is shown in the sketch. The steam passage leads from E to the passage beyond F, and two discs A, B, are fitted on the spindle. The pressure on the under side of B balances that on the upper side of A, whereby the steam is powerless to resist the raising of the compound valve.

Fig. 160.

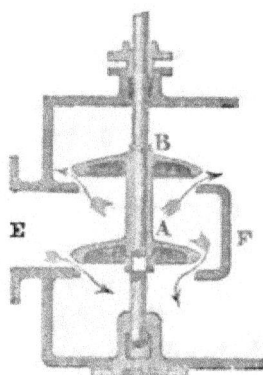

The student should compare this valve with the *crown valve* described in *Art.* 113, and he will see that the principle of both valves is the same ; the opposition of fluid pressure taking place on the curved surface of the crown valve, but directly on the discs of the valve here described. In organs a double disc valve is used ; but the construction is different, for the valves are attached to the two ends of a lever, centred midway between the openings, whereby one moves inwards against the air pressure, and the other moves outwards in the direction of the current.

CHAPTER IX.

ON MOTION IN ONE PLANE.

156. In the Introduction we proved four equations which apply in the case of a body falling from rest, viz.,

$$v = gt,\ s = \frac{1}{2} t v,\ s = \frac{1}{2} g t^2,\ v^2 = 2 g s,$$

and we also pointed out that these equations, though proved only in the case of a body falling under the action of gravity, were not so restricted, but were true generally.

That is, if f be the velocity generated in one second in a body by a constant force acting in the line of its motion, we have the equations,

$$v = ft, \quad s = \frac{1}{2} t v, \quad s = \frac{1}{2} f t^2, \quad v^2 = 2fs.$$

Prop. To find the equations of motion when a body is projected in a given direction and acted on by a *constant force* in the line of its motion.

The velocity of projection would carry the body uniformly through space, and if we conceive that a velocity equal to that of projection is impressed both on the body and the region in which it is, the *relative motion* of the body to the surrounding objects will be unaltered. But in that case the body will be at rest, and the region around it will move.

Now suppose the force to act on the body at rest, let f be the velocity which it generates in one second, t the time of motion, then ft is the velocity acquired by the body in t seconds, and $\frac{1}{2} f t^2$ is the space described in the same time.

Let v be the velocity of projection, then the region has been moving with a uniform velocity v, and has described a space t v in t seconds. Hence if v be the velocity, and s the space actually described by the body relatively to fixed objects in the region during the time t, we have

$$v = \text{v} + ft, \quad \text{or } v = \text{v} - ft, \quad \ldots \ldots \quad (1)$$

$$s = t\text{v} + \frac{1}{2} f t^2, \quad \text{or } s = t\text{v} - \frac{1}{2} f t^2 \quad \ldots \quad (2)$$

according as the force acts to increase or destroy the velocity of projection.

In like manner, we have $v^2 = (\text{v} + ft)^2$

$$= \text{v}^2 + 2\text{v}ft + f^2 t^2,$$

$$\therefore v^2 = \text{v}^2 + 2fs, \quad \text{or } v^2 = \text{v}^2 - 2fs, \quad \ldots \quad (3)$$

taking the signs as in equations (1) and (2).

Note. In the case of a body projected vertically we have merely to write g in the place of f, taking the negative signs when the body is thrown upwards, and the positive signs when it is thrown downwards.

Ex. 1. The velocity of projection of a body is $8g$: find the time in which it rises vertically through $14g$.

Here $v^2 = u^2 - 2gs = 64g^2 - 2g \times 14g = 36g^2$. $\therefore v = \pm 6g$.

and $6g = 8g - gt$, $\therefore t = 2$ seconds.

Ex. 2. A body is projected vertically upwards with a velocity of 100 feet per second : find its height after 3 seconds. *Ans.* 155·1 feet.

Ex. 3. With the same data, after what interval is it 140 feet above the point of projection ? *Ans.* After 2·13 or 4·08 seconds.

Ex. 4. A body is thrown vertically upwards with a velocity of 161 feet per second from the top of a tower $214\frac{2}{3}$ feet high. In what time will it reach the ground, and what velocity will it acquire?

Ans. Time = 11·2 seconds, velocity = 199·4 feet.

Ex. 5. A body thrown up with a velocity of 60 feet reaches the top of a tower in 2 seconds. Find the height of the tower. *Ans.* 55·6 feet.

Ex. 6. Two bodies, A and B are moving towards each other, from the two extremities of a vertical line of length a, A having been let fall from rest at the top of the line at the same instant that B was projected upwards from the bottom with a vertical velocity $\frac{3a}{2}$. Determine where they will meet. *Ans.* 7·155 feet from the top of the line.

ON MOTION ON AN INCLINED PLANE.

157. If a body of weight w be placed on a smooth plane inclined at an angle a to the horizon, the component of w along the plane is w sin a. Hence the velocity generated in one second in the body by this component will be g sin a. If, therefore, we write g sin a for g in the previous equations we shall obtain formulæ applicable to motion on an inclined plane, and subject only to the restriction that the body is so projected as to move always in the same vertical plane.

Hence, with the same notation as before, we have

$$v = v \pm tg \sin a, \quad . \quad . \quad . \quad . \quad (1)$$

$$s = tv \pm \frac{1}{2}gt^2 \sin a, \quad . \quad . \quad . \quad (2)$$

$$v^2 = v^2 \pm 2gs \sin a \quad . \quad . \quad . \quad (3)$$

Ex. 1. A body slides down a smooth incline in 10 seconds, and acquires a velocity of $5g$, show that height : length = 1 : 2.

Ex. 2. A rough plane rises 4 feet in 3 horizontal, the co-efficient of

friction is $\frac{1}{3}$, and a body is projected up the plane with a velocity $3g$.
Find how far it moves along the plane, and the time before it returns
to the starting point. *Ans.* $4\frac{1}{2}g$ feet, $5\cdot7$ seconds nearly.

Ex. 3. A body is projected down an inclined plane with a velocity
acquired in falling down its height, and it describes the length of the
plane in the time of falling down its height. Find the elevation of the
plane. *Ans.* The angle whose sine is $\sqrt{2}-1$.

Ex. 4. A weight P, descending vertically, draws W up an inclined
plane, whose angle of elevation is 30°. Determine the velocity of P
after n seconds have elapsed.

ON MOTION IN A CURVE WHOSE PLANE IS VERTICAL.

158. The velocity which a body acquires in falling down a
smooth groove or curve when its weight is the only force
causing motion, may be found very simply from the doctrine
of work. The reaction of the curve is everywhere perpendicular to the direction of motion, and it is clear that this
reaction does no work in accelerating or retarding the velocity of the body. It is the pull of the earth that does work
and impresses a velocity equal to that acquired in falling
down the vertical depth of the curve. Hence if h be the
vertical depth of the curve, v the velocity of projection, v
the velocity of the body, we have, as before,

$$v^2 = v^2 + 2gh, \quad \text{or } v^2 = v^2 - 2gh.$$

THE VELOCITY OF EFFLUX OF WATER UNDER PRESSURE.

159. Connected with this subject of the motion of a falling
body is the law which regulates the velocity with which water
issues from a small orifice in the side of a vessel containing
it. The fact here presented is one of great interest. It is
a property of fluids under pressure that their particles are
ready in an instant to start out with a velocity dependent on
the pressure, and to take up motion at once in any path
which may be opened for them. In doing so they obey
definite laws which are only partially understood. The conversion of fluid pressure into momentum, and the reconversion of such momentum into pressure, should be carefully

studied wherever the phenomena are observed. We have here only space for one simple proposition, restricted in its application to *small* jets.

Prop. To find the velocity with which water issues through a very small orifice in a vessel containing it.

For simplicity we will suppose the vessel to be supplied with water as it leaks out, whereby the surface remains at one height. Let *v* be the velocity at the orifice, *w* the weight of a small portion of the fluid issuing with that velocity, and *h* the depth of the orifice below the surface.

Then the kinetic energy stored up in *w* is represented by $\dfrac{w v^2}{2 g}$. But the portion *w*, if it had descended from the surface to the orifice, would have had an amount of work represented by *w h* impressed upon it; and if we further suppose that there is no loss by friction, we may equate these two expressions for the work done, and shall have

$$\frac{w v^2}{2 g} = w h,$$

or $v^2 = 2 g h$, and $v = \sqrt{2 g h} = 8 \sqrt{h}$, very nearly.

It may be objected that the motion begins long before any portion of the fluid at the surface has reached the orifice, which is quite true, and the proof is not a complete demonstration. It is, however, certain that the flow goes on without change so long as *h* remains constant, and that after a time some portion of the liquid near the surface will have escaped from the orifice, according to our hypothesis. The safer course probably is to regard the formula as the result of experiment.

There is an instructive illustration by Mr. Bramwell, who caused a jet of water to issue from a cistern and to impinge against a horizontal mouth-piece at the bottom of a vertical glass tube. The water in the cistern was maintained at a constant height of 4 feet, and the water in the glass tube rose to a height of 3 feet $11\frac{3}{4}$ inches, and there remained. Here was an example of the reconversion of the momentum of the liquid impinging at the mouth-piece into the pressure of the head supported in the glass tube, the loss being represented by a head of $\frac{1}{4}$ of an inch.

In the proposition given above the pressure on the surface of the water is equal to that on the issuing jet. It may happen that these pressures are very unequal, as is the case when water enters the condenser of a steam-engine ; we then take their difference and obtain an effective head of water by combining the imaginary head due to this difference with the actual head in the vessel.

Ex. 1. The head of water is 3 feet, the external barometer stands at 30 inches, and the barometer gauge of a condenser shows 24 inches : find the velocity with which water enters the condenser through a small orifice.

Here the artificial head = 24 × 13·6 inches = 27·2 feet.

Therefore $v^2 = 2g(3 + 27.2) = 64.4 \times 30.2 = 1944.88$ feet,

and $v = 44.1$ feet per second.

Ex. 2. The reconversion of liquid momentum into pressure is also exhibited by the experiments of Mr. Ramsbottom when arranging for the supply of water to a locomotive while running.

Let $h = 7\frac{1}{2}$ feet, then $v^2 = 64.4 \times 7.5 = 504$ feet, and $v = 22.5$ feet per second nearly, which corresponds to about 15 miles per hour. It was found, on trial, that with a velocity of 15 miles per hour, the water was raised $7\frac{1}{2}$ feet up the delivery pipe, and remained stationary at that height. The velocity of the water relatively to the delivery pipe was fifteen miles an hour, although in one sense the water was at rest.

THE VELOCITY OF EFFLUX OF GASES.

160. In estimating *approximately* the flow of gases through a small orifice it is usual to make the same hypothesis as that adopted in finding the height of the homogeneous atmosphere, viz., that the gas is incompressible and behaves as a liquid.

Let the pressure of air support 1 inch of water. We know that 29·922 × 13·596 inches of water balance 26,214 feet of incompressible air, therefore 1 inch of water will balance 64·4 feet of air. This fact will be very easily remembered.

Or we may say that the density of air at 32° F. is ·380728 lbs. per cub. foot, and that of water at 39°·4 F. is 62·425 lbs. per cub. foot, from which data we deduce the same result.

We will suppose the air to be contained in a vessel A and to flow into a vessel B. Let the difference of pressures in A and B as measured by a water gauge be x inches. Then the velocity v of issue is approximately given by the formula

$$v = \sqrt{2g \times 64\cdot4\ x} = 64.4 \sqrt{x}$$

The general formula is $v = \sqrt{2gh}$, where h is the height of a column of air or of the gas, supposed incompressible, whose weight balances the difference of pressures outside and inside the vessel containing it.

Ex. 1. Steam in a boiler is at 60 lbs. actual pressure; compare the velocities with which steam and water will issue from small orifices respectively connected with the steam and water spaces.

At a pressure of 60 lbs. the volume of steam is 464 times that of the water from which it is produced. Both the steam and the water rush out under a pressure equivalent to 3 atmospheres, and the *head* of steam is 464 times the *head* of water, whatever that may be, therefore

vel. of steam : vel. of water $= \sqrt{464} : 1 = 21\cdot5 : 1$.

AN EXPANDING CHIMNEY FOR THE DISCHARGE OF AIR.

161. In Fig. 19 the air discharged from the fan is not poured directly into the external atmosphere, but passes through an expanding chimney, which gradually reduces its velocity, the object being to prevent a loss of power. It is evident that the velocity of a stream of air when passing through a pipe will become less as the pipe is enlarged. Of that there can be no doubt. Taking with us the principle that action and reaction are equal and opposite, let us consider the case of a mass of air entering the narrow neck of an expanding *flue* or *chimney* at a high velocity, and discharging itself finally at a low velocity. This illustrates the reconversion of momentum into pressure. The molecules of air move more and more slowly and finally encounter the inert atmosphere outside. Here they are crowded together, the space becomes more densely packed, and finally this reduced momentum terminates quietly in an increase of air pressure. If the air rushed out, unaided by the expanding chimney, it would meet with much greater resistance; it would set up eddies and would be clogged in every direction, whereby the engine would be more severely taxed, and steam-power would be wasted.

THE PARALLELOGRAM OF VELOCITIES.

162. It is evident that the velocity of a body at any instant may be completely represented by a straight line.

1. The length of a line represents the number of feet travelled over by the body when moving uniformly, or it represents the number of feet which *would be described* in one second if the body retained for one second the velocity which it has at the instant considered.

2. The direction of the line represents the direction of motion.

Now the proposition of the parallelogram of forces applies to the composition of forces represented by straight lines, and it would apply equally to the composition of velocities represented by straight lines. *Mutatis mutandis* the proof may be gone over almost word for word and the conclusion is the same. Hence *we may resolve or compound velocities just as we resolve or compound forces.* Referring to Fig. 33, we say that a body starting from A with a velocity v in a direction making an angle a with A X has a velocity $v \cos a$ along A X and $v \sin a$ perpendicular to it, and thus we resolve v into its components.

ON THE MOTION OF PROJECTILES.

163. We have only space for a few propositions in connection with this subject, and we shall suppose the motion to occur in a space devoid of air. It is a well-established fact that the resistance of the air exerts a most important influence in modifying all theoretical conclusions as to the motion of a projectile, and accordingly the results we are about to deduce cannot be applied in practice. Nevertheless, those who intend to pursue the science of gunnery must begin by studying the path of a projectile *in vacuo.*

Prop. *A body projected in any direction which is not vertical, and acted on by gravity, will describe a parabola.*

Let A be the point of projection, A T the direction, and *v* the velocity of projection. Conceive that the body moves in some unknown curve A P, the nature

FIG. 161.

of which we are about to determine, and let *t* be the time of flight from A to P. Draw P T vertical and meeting A T in T, complete the parallelogram T V. It is evident that the body has two motions always existing, viz., (1) the original motion of projection in A T, (2) the motion which it would acquire in falling from rest during the time of flight. These two rectilinear motions combine to give the curved path A P.

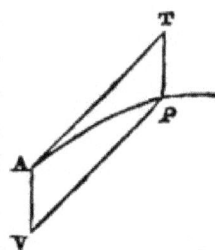

Therefore $AT = tv$, $TP = \dfrac{1}{2} g t^2$,

$$\therefore\ TP = \frac{1}{2} g \cdot \frac{AT^2}{v^2}, \quad \text{or } PV^2 = \frac{2v^2}{g} \times AV.$$

This relation between P V and A V indicates that the curve is a *parabola* whose axis is vertical.

Some miscellaneous questions now present themselves.

1. *To find the time of flight on a horizontal plane* A D *drawn through the point of projection* A.

Let *v* be the velocity of projection, α the angle of elevation. Then $v \cos α$, $v \sin α$, are the resolved velocities in and perpendicular to A D.

During the flight, $v \sin α$ is destroyed by the pull of the earth, and generated again, for the body comes down at D with the velocity with which it rose from A.

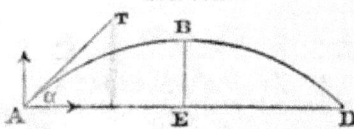

FIG. 162.

Let t be the time of flight, then in time $\dfrac{t}{2}$ the velocity $v\sin a$ is destroyed by gravity, therefore

$$v\sin a = \frac{gt}{2}, \quad \text{or} \quad t = \frac{2v\sin a}{g}.$$

2. *To find the range* A D.

Since the horizontal velocity, $v\cos a$, is uniform, we have

$$\text{A D} = t v\cos a = v\cos a \times \frac{2v\sin a}{g} = \frac{2v^2}{g}\sin a\cos a.$$

3. *To find the greatest height to which the body rises.*

Since $v\sin a$ is generated in falling down a height equal to B E, we have

$$v^2\sin^2 a = 2g \times \text{B E}, \therefore \text{B E} = \frac{v^2\sin^2 a}{2g}.$$

4. *To find the time of flight on an inclined plane* A D *passing through the point of projection.*

Let A be the point of projection, A D the plane inclined at an angle a to the horizon, A T the direction of projection making an angle θ with the plane,

FIG. 163.

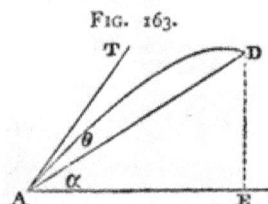

v the velocity of projection, and let t be the time of flight through A D.

Then $v\cos\theta$, $v\sin\theta$, are the components of v along A D and perpendicular to it; also $g\sin a$, $g\cos a$ are the components of g along D A and perpendicular to it.

Now the velocity $v\sin\theta$ is destroyed in time $\dfrac{t}{2}$ by the resolved pull of the earth in a direction perpendicular to A D,

therefore $\qquad v\sin\theta = \dfrac{t}{2}g\cos a$, and $t = \dfrac{2v\sin\theta}{g\cos a}$.

5. *To find the range* A D.

Since the velocity $v\cos\theta$ is affected only by the resolved pull of the earth along D A, we have

$$\text{A D} = t\,v\cos\theta - \frac{1}{2}g\sin a.\,t^2,$$

$$= \frac{2\,v^2\sin\theta\cos\theta}{g\cos a} - \frac{1}{2}g\sin a.\frac{4\,v^2\sin^2\theta}{g^2\cos^2 a},$$

$$= \frac{2\,v^2.\sin\theta}{g\cos^2 a}\left(\cos a\cos\theta - \sin a\sin\theta\right),$$

$$= \frac{2\,v^2.\sin\theta.\cos(a+\theta)}{g\cos^2 a}.$$

Aliter. Draw D E perpendicular to A E, then A E would be described in time t with a uniform velocity $v\cos(a+\theta)$,

$$\therefore \text{A E} = t\,v\cos(a+\theta) = \frac{2\,v^2\sin\theta\cos(a+\theta)}{g\cos a},$$

But $\quad$ A D $= \dfrac{\text{A E}}{\cos a}\quad \therefore$ A D $= \dfrac{2\,v^2\sin\theta\cos(a+\theta)}{g\cos^2 a}.$

6. *To find the greatest perpendicular distance of the body from the inclined plane.*

The body will be moving parallel to the plane when most distant from it, for $v\sin\theta$ is then destroyed.

Let s be the required distance, then, as in **case 3,** we have

$$v^2\sin^2\theta = 2\,g\,s\cos a \quad\therefore\quad s = \frac{v^2\sin^2\theta}{2\,g\cos a}.$$

7. *To find a relation between the horizontal and vertical co-ordinates of the projectile at any instant.*

FIG. 164.

Let A be the point of projection, A T the direction of projection, A X, A Y, horizontal and vertical axes, P the position of the projectile at the end of t seconds, v the velocity of projection.

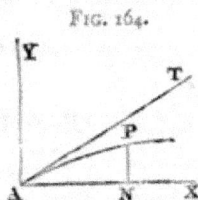

Then, as before, $x = t\,v\cos a,\ y = t\,v\sin a - \dfrac{1}{2}g\,t^2,$

Substituting for t, we have $y = x\tan a - \dfrac{g\,x^2}{2\,v^2\cos^2 a}.$

EXAMPLES ON THE MOTION OF PROJECTILES.

164. In *Art.* 135 we explained the meaning of the term *elasticity*, and we may here point out that when the velocity of rebound of a body impinging in a perpendicular direction upon a plane, is equal to the velocity with which it strikes, the elasticity between the body and plane is said to be perfect ; if the velocity of rebound were $\frac{1}{2}$ that of before impact, the elasticity would be ·5, and so on. It is usual to call the ratio of these two velocities e, and many exercises are given on this subject. According to the present view of the nature of elacticity, the co-efficient e is equal to $\frac{15}{16}$ for hardened steel and $\frac{5}{9}$ for glass. Some of the following examples will illustrate these remarks.

Ex. 1. A ball of elasticity e, is projected with a velocity v, at an elevation a, and at a distance a from a vertical wall. Prove that it will return to the point of projection when $v^2 \sin 2\,a = \dfrac{ga\,(1 + e)}{e}$.

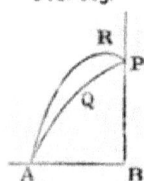

Fig. 165.   Let $AB = a$, then the whole time of flight is equal to the sum of the times through AQP and PRA.

$$\therefore \frac{2v\sin a}{g} = \frac{a}{v\cos a} + \frac{a}{ev\cos a} = \frac{a}{v\cos a}\cdot\left(\frac{1 + e}{e}\right),$$

$$\therefore v^2 \sin 2a = \frac{ga\,(1 + e)}{e}.$$

Ex. 2. A perfectly elastic ball falls from a height h upon a plane inclined at 30° to the horizon. If t be the time of falling, show that the time of flight $= 2t$, and the range $= 4h$.

Here the resolved velocities are $\dfrac{v}{2}$, $\dfrac{v\sqrt{3}}{2}$, where $v^2 = 2gh$.

Let t' be the time of flight, then $\dfrac{v\sqrt{3}}{2} = g\dfrac{\sqrt{3}}{2}\cdot\dfrac{t'}{2}$ or, $t' = \dfrac{2v}{g} = 2t$.

Also range $= \dfrac{t'v}{2} + \dfrac{1}{2}\cdot\dfrac{g}{2}\cdot t'^2 = \dfrac{2v^2}{g} = 4h$.

If the elasticity were $\dfrac{1}{2}$, the range would be $\dfrac{3h}{2}$.

Ex. 3. Find the elevation of a projectile which has the greatest range on a horizontal plane.

Here the range $= \dfrac{2v^2}{g} \sin a \, \cos a = \dfrac{v^2}{g} \sin 2 a,$

and the range is greatest when sin 2 a is greatest, or when **a = 45°.**

The greatest range ever obtained by a projectile has been 11,243 yards. The projectile weighed 250 lbs., and it was fired from a Whitworth 9-inch gun at an **elevation** of 33°. The trial took place at Shoeburyness in 1868.

Ex. 4. The velocity **of projection** of a projectile **is 1,000 feet per** second, and **the range is 500** yards: find the angle of elevation, **and the** greatest height **to which it** rises above the horizontal plane.

Ans. 1° 23', about 9 feet.

Ex. 5. **A body is projected** at 45° with a velocity acquired in falling down 2⅔ the height of a tower, viz. *h.* Within **what** limits of distance from the tower will the projectile pass over it ? *Ans.* 4 *h*, or $\dfrac{4\,h}{3}$.

Ex. 6. **Find** the two angles of elevation which give **the** same range **on a horizontal** plane through the point of projection.

Ans. The angles are a, and 90 − a.

Ex. 7. If *v*, *v'*, *v''*, be the velocities at three points P, Q, R, in the path of a projectile where the inclinations to the horizon are a, a − β, **a − 2** β, respectively, and *t, t'*, be the **times of** describing P Q, Q R, respectively, show that $v'' t = v t'$, and $\dfrac{1}{v} + \dfrac{1}{v''} = \dfrac{2 \cos \beta}{v'}$.

Here $v \cos a = v' \cos (a - \beta) = v'' \cos (a - 2\beta),$

$g t = v \sin a - v' \sin (a - \beta),$

$g t' = v' \sin (a - \beta) - v'' \sin (a - 2\beta).$

The solution is merely an exercise in combining these equations.

CHAPTER X.

ON CIRCULAR MOTION.

165. *Prop.* A body of weight w, describes a circle of radius **r,** with a *uniform* velocity *v* ; to find the direction and magnitude of the force F which produces this motion.

It is clear that some force must act, otherwise the body would describe **a straight line and** not a circle. Again, the force in action **neither accelerates nor** retards the body, and therefore its direction must so change as to be always perpendicular to the line in which the body is moving, and must point to the centre of the circle.

Let A P be a very small arc of the circle described round C in time *t*, draw C P Q meeting the tangent at A in Q.

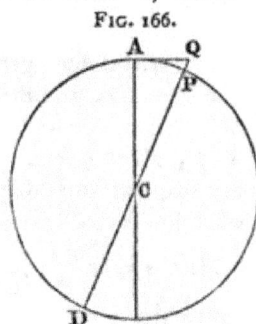

Fig. 166.

If no force acted, the body would move from A to Q in a small interval of time, but the force F pulls it through Q P, and retains it in the circle. Let *f* be the velocity generated in 1 second in the body of weight W by the action of the force F, then

$$F = \frac{W f}{g}.$$

Produce Q P C to D, then by a property of the circle, Q P × Q D = Q A ^{2}. Let Q P = *x*, therefore

$$x\,(2\,r + x) = Q A^2, \text{ or } 2\,r\,x + x^2 = Q A^2.$$

Now the arc A P is so small that we may take A P as equal to Q A, and may neglect x^2, ∴ $2\,r\,x = $ A P ^{2}.

But the motion in Q P is quite independent of that in the circle, therefore Q P $= \frac{1}{2}\,f\,t^2$., or $x = \frac{1}{2}\,f\,t^2$. Also the motion in the circle is uniform, therefore A P $= t\,v$.

$$\therefore 2\,r \times \frac{1}{2}\,f\,t^2 = t^2\,v^2, \text{ and } f\,r = v^2,$$

$$\therefore F = \frac{W v^2}{g\,r}.$$

This is the equation which governs the law of circular motion ; in applying it we observe

1. That an *inward pull* must act on the body, which is called the *centripetal* or centre-seeking force.

2. That the reaction to this inward pull will be felt on the centre when the body is attached thereto by a string or light rod. The reaction produces an *outward pull* on the centre itself, which is called the *centrifugal* or centre-flying force.

The result is that the centre continually tends to move in a direction pointing to the revolving body. When a wheel is loaded on one side, so that its centre of gravity does not exactly coincide with the centre of figure, there will be a tendency to jump in the bearings during each revolution.

Suppose a wheel 3 feet 6 inches diameter to run at a velocity of 50 miles an hour, and to be a little out of truth, whereby 9 lbs. in excess is distributed on one part of the rim.

$$\text{Here } w = 9 \text{ lbs}, \ v = \frac{220}{3}, \ r = \frac{7}{4},$$

$$\therefore \frac{w\,v^2}{g\,r} = \frac{9 \times 220 \times 220}{9 \times 32\cdot2 \times \frac{7}{4}} = \frac{96800}{112\cdot7} = 858\cdot8 \text{ pounds.}$$

This is a pull of more than $7\frac{1}{2}$ cwts., and each revolution occupies about $\frac{1}{13}$ second. Every time that the loaded part of the wheel describes the upper half of its circular path this force of $7\frac{1}{2}$ cwts. is in action to lift the wheel, and is a direct lifting force at the instant of passing the highest point. The vibration, which is set up in heavy revolving mechanism when unbalanced, soon becomes intolerable. Yet in the early days of screw engines the two cranks were unbalanced. The same is true in light mechanism, running at a high velocity; thus in some wood-carving machinery a tool was fixed in the rotating spindle by a small capstan head-screw, weighing 150 grs.; and although about half that weight was buried in the spindle, it was found necessary to balance the screw-head, as an injurious vibration was set up. Here the spindle made 7,000 revolutions per minute.

Ex. 1. A body whose weight is 10 lbs. is whirled round in a circle of 10 feet radius with a velocity of 30 feet per second. Find the force F to the centre.

$$\text{Here } F = \frac{w\,v^2}{g\,r} = \frac{10 \times 900}{32\cdot2 \times 10} = 28 \text{ lbs nearly.}$$

Ex. 2. The diameter of a circle is 10 feet; find the time of a revolution when $F = w$. *Ans.* 2·445 seconds.

Ex. 3. A body, of weight w, is whirled round by a string in a

S

vertical circle. Prove that the string must be able to support at least 6 times the weight of the body.

The velocity at the highest point must be just enough to keep the body in the circle to prevent its dropping out of the curve. Let v be this velocity, a = radius of circle, then $v^2 = g\,a$.

Also (velocity)^2 at lowest point $= 2g \times 2a + v^2 = 5g\,a$.

$$\text{Hence tension of string}: \text{w} = g + \frac{5g\,a}{a} : g = 6 : 1.$$

$$\therefore \text{ Tension of string} = 6 \text{ w}.$$

MEANING OF THE TERM VIS VIVA.

166. It has been explained that the amount of work stored up in a body of weight w, when moving with a velocity v, is measured by the expression $\dfrac{\text{w}\,v^2}{2g}$.

Formerly it was the practice to call this quantity $\frac{1}{2}$ the *vis viva* of the body, and the term *vis viva* is still very commonly used, although beginning to be replaced by the phrase *kinetic energy*. It will be understood that the *vis viva* of a body is twice its *kinetic energy*.

When a body is in motion in any line, whether straight or curved, and has a given linear velocity, we estimate the work stored in it by $\frac{1}{2}$ the product of the mass into the square of its velocity. Thus the work stored up in a heavy weight placed at the end of a revolving bar depends only on the mass and the square of the linear velocity. The force to the centre in no way influences the result, except so far that it is necessary to maintain the motion; and we shall illustrate these remarks by referring to a stamping press.

THE FLY PRESS.

167. The Fly Press is employed for stamping or coining metals. Here two massive balls are fixed at the ends of a long bar or lever, connected with a screw of somewhat rapid pitch, which carries the die or punch intended to act on the piece of metal. A workman rotates the screw until the work is reached; the whole moving mass is then brought suddenly to rest, and a force of great intensity impresses the die.

Let w be the combined weight of the two balls, v the velocity of either of them at the instant of impact, then $\dfrac{\text{w}\,v^2}{2\,g}$ is the work stored up in them. Also, let R be the resistance overcome, *supposed to be constant*, and y the space moved through by the end of the screw, then R y is the work done against R. Hence, by the principle of work,

$$\text{R}\,y = \frac{\text{w}\,v^2}{2\,g}, \quad \text{or} \quad \text{R} = \frac{\text{w}\,v^2}{2\,g\,y}.$$

Ex. Two balls, each weighing 100 lbs., are placed at the ends of a horizontal bar or lever 5 feet from the centre of motion. The lever imparts motion to a vertical screw of 2 inches pitch working a punch, as in the ordinary punching-press. What resistance will the punch overcome if the balls have a velocity of 10 feet per second at the moment of impact, and the punch is brought to rest after traversing a distance of $\frac{1}{10}$ of an inch?

(Science Exam. 1873.)

Here $v = 10$, W $= 100 + 100 = 200$.

$$\therefore \frac{\text{w}\,v^2}{2\,g} = \frac{200 \times 100}{64\cdot4} = 310\cdot5 \text{ foot pounds.}$$

$$\text{R} = \frac{\text{w}\,v^2}{2\,g\,y} = \frac{310\cdot5}{\frac{1}{120}} = 37260 \text{ pounds.}$$

Neither the length of the lever nor the pitch of the screw have anything to do with the answer to this example where v is given.

168. It being understood that uniform circular motion is impossible unless the body be pulled in towards the centre by a force, and there being no action without an equal and opposite reaction, it is evident that we can imitate in a body at rest the conditions which obtain during circular motion by supplying a force equal and opposite to this centre-seeking force.

For example, drop a marble into a hollow circular cylinder, and set the cylinder in rotation about its axis, which we assume to be vertical. The marble will run to the side, and press against it. The side will react on the marble, and give the force necessary for the circular motion. If we wished to imitate in the cylinder when at rest the action which is going on during the rotation, we should merely press the marble against the side of the vessel with a force equal to that which before kept up the circular motion. The result is, that when the body is revolving we have a force *tending towards the centre*; when it is at rest, we imitate the state of things by means of a force equal to the former, and tending *from the centre outwards.*

In order to make this matter still more clear, we may discuss the action of Watt's conical governor, or the *conical pendulum,* as it is often termed.

THE PRINCIPLE OF THE CONICAL PENDULUM.

169. It is a well-known fact that a body suspended by a long string may be set in motion so as to describe for some time a circular path in a horizontal plane.

Let D represent a body of weight w, suspended at c by the string c d, and describing a horizontal circle of radius D B with a uniform velocity v. Let $c B = h$, $B D = r$, also let T be the tension of the string D c, and t the time of a revolution. Applying the method just explained we shall suppose the body D to be at rest, and to be acted on by three forces, viz. (1) its weight, (2) the tension T, and (3) the force $\frac{w v^2}{g r}$ *acting from the centre outwards.*

FIG. 167.

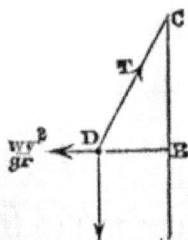

Since D is at rest on this hypothesis, we have

$$r : h = \frac{w v^2}{g r} : w = \frac{v^2}{g r} : 1. \quad \therefore v^2 h = g r^2.$$

Also the motion is uniform; therefore $2 \pi r = t v$, and

$$t = \frac{2 \pi r}{v} = 2 \pi \sqrt{\frac{h}{g}}.$$

Hence *the time of a revolution varies directly as the* square
root of the height of the cone.

Cor. 1. If ω be the angular velocity of the line B D, we
have $v = \omega r$; $\therefore \omega^2 r^2 h = g r^2$, and $\omega^2 h = g$.

Cor. 2. If n be the number of revolutions per minute, we
have $t = \dfrac{60}{n}$, and $n = \dfrac{30}{\pi} \sqrt{\dfrac{g}{h}}$.

170. We have next to show that if ω be increased, the
ball D will fly off to a greater distance
from C B. Let $CD = l$, $DCB = \theta$, then
$h = l \cos \theta$, and $\omega^2 h = g$.

Fig. 168.

$\therefore w^2 l \cos \theta = g$, or $\cos \theta = \dfrac{g}{l \omega^2}$.

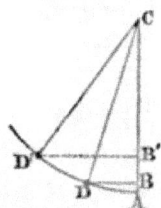

If ω be increased, $\cos \theta$ is diminished,
and θ is increased, whereby D moves up
into a new position, such as D′. As a general rule, therefore, the body takes a low position with a moderate velocity,
and rises higher as the velocity increases.

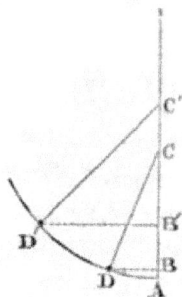

If the body be constrained to move in a parabola, this
result will be modified in a remarkable manner, for D will
remain on the curve at *one velocity only, and will take any
position, whether high or low.* The only
condition affecting C D is that it shall be
perpendicular to the curve in every position ; and it is the property of a parabola
that if D′c, D′ c′ be perpendiculars to the
curve at D and D′, and D B, D′ B′ be horizontal lines, meeting the vertical axis in
B, B′, the height of the cone C B is equal
to the height of the cone c′ B′. Hence,
with a suitable velocity, the body would
be at equilibrium at D′ as well as at D, and the same is true
for any other point of the curve.

Fig. 169.

Also, $\omega^2 \times CB = g$, and when C B is given, ω has only one
numerical value ; therefore only *one* velocity is possible. We
thus see what can be done with a parabolic pendulum.

ON THE ROTATION OF LIQUIDS IN AN OPEN VESSEL.

171. We have already pointed out that when a vessel containing water is made to revolve about a vertical axis, the water assumes a hollow cup-like form, and we propose now to investigate the law which determines the nature of the surface.

If we place some fine sand in a shallow cylindrical vessel open at the mouth, and rotate the vessel rapidly about its axis, which must be vertical, the sand will pile itself against the side, but its surface will not be curved. The slope will be nearly a straight line. If, in the place of sand, we pour some water or mercury into the vessel and rotate it as before, the liquid will assume a beautiful curved shape. On trying the experiment with clean mercury, and operating very carefully, it will be possible to obtain a concave reflecting surface competent to form an image of any brilliant object, such as the carbon points of the electric lamp, and to define it fairly on the ceiling of the room. The surface so obtained would be generated by the revolution of a parabola about its axis, and is called a *paraboloid of revolution.*

Since a liquid is an assemblage of molecules, the mechanical conditions applicable to a liquid rotating with an angular velocity ω, are precisely the same as those we have investigated. We suppose the liquid to be at rest, and apply a force $\dfrac{m\,\omega^2\,r}{g}$ to each molecule of weight m, and whose distance from the axis of rotation is r. Since the rotation introduces horizontal forces only, the liquid will in other respects be subject to the ordinary laws of fluid pressure. Our object is to find the form of the surface, and we argue that its direction at any point is determined by the fact that it must be perpendicular to the forces acting upon it at that point.

172. *Prop.* To find the form of the surface of a liquid rotating uniformly about a vertical axis.

We shall suppose the liquid to be placed in a circular

cylindrical vessel D B C E, whose axis K A is the axis óf rotation.   Let ω be the angular velocity of rotation, draw P G
perpendicular to the surface from any point P, and P N perpendicular to A G.   Then the forces

FIG. 170.

acting on a particle of weight m at P
are (1) its weight, (2) $\dfrac{m\,\omega^2\,P\,N}{g}$,

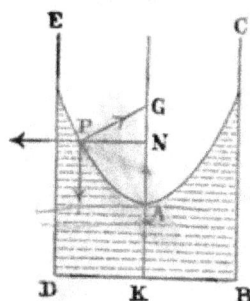

(3) the fluid pressure at P; also force
(3) must be the resultant of forces
(1) and (2), and must act in P G. But
P is at rest, therefore the forces are
proportional to the sides of the
triangle P G N, and we have

$$G\,N : P\,N = m : \frac{m\,\omega^2\,P\,N}{g} \quad \therefore\; G\,N = \frac{g}{\omega^2}.$$

Hence G N is constant.   But this, as we have already seen,
is the property of a parabola, and we return to the parabolic conical pendulum, as was inevitable, since P has no
tendency to move along the curve A P.   It is always instructive to observe the same law under two different aspects.

The conclusion, then, is that the surface of the liquid is a
paraboloid of revolution.   If it were possible to rotate a
large vessel of mercury without any vibration, we should
obtain the most perfect concave reflector for a telescope,
which would be restricted to viewing stars near the zenith.
The mechanical difficulties to be overcome before such a
telescope could be constructed are very formidable.   It is
essential (1) to keep the rotation of the liquid perfectly
constant, as otherwise the focal length of the mirror will
change.   (2) There must be no vibration, for the slightest
tremor on the surface of the mercury would break up the
image of a star just as the rippling of a river breaks up the
reflected image of a distant light.

173. Prop. *When a mass of liquid rotates in a cylindrical
cup, the vertex A falls as much below the original level as the
edge E rises above it.*

Repeating the previous figure and notation, let efc be the
original level of the liquid, which, on rotation, just rises to E and falls to A.

FIG. 171.

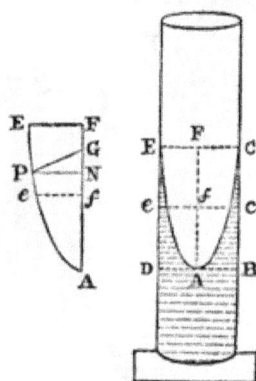

Then it is a property of the paraboloid that its volume is $\frac{1}{2}$ that of the circumscribing cylinder, hence

volume $E A C = \frac{1}{2} \pi A D^2 \times F A$,

$$= \pi A D^2 \times A f, \text{ also,}$$

since $A f$ was the depth of the liquid before rotation, therefore

$$F A = 2 A f,$$

or the point f bisects $A F$, which proves the proposition.

RAMSBOTTOM'S VELOCIMETER.

174. When experiments were in progress in order to ascertain the mechanical conditions for taking up water from a trough into the tender, it was desirable to know by inspection the rate at which the engine was running. Mr. Ramsbottom employed for this purpose a glass cylinder half full of oil, and set it in rotation by a cord passing round the trailing axle of the engine. The vertex of the surface sank more and more as the rotation increased, and its position indicated at a glance the speed of the locomotive.

It is a property of the parabola that $P N^2 = 2 G N \times A N$.

When the velocimeter rotates with an angular velocity ω, let the vertex be depressed through a depth x, and we have

$$E F^2 = 2 G N \times 2x = \frac{4 g x}{\omega^2}.$$

But $E F$ is constant, therefore x varies as ω^2, or the depression of the surface varies as the square of the angular velocity. Thus the vessel may be graduated.

Ex. 1. A cup 2 inches diameter and 6 inches deep, is half filled with oil ; find the number of revolutions per minute, when the oil is just brought up to the point of overflow.

Ex. 2. A cylinder of radius a and depth b is filled with water, and made to revolve with an angular velocity ω; to find the quantity of water thrown out.

Let x be the depth to the vertex, which we will suppose to be less than b. Then volume thrown out

$$= \frac{\pi a^2 x}{2} = \frac{\pi a^2}{2} \cdot \frac{\omega^2 a^2}{2g} = \frac{\pi a^4 \omega^2}{4g}.$$

If half be thrown out, $\dfrac{\pi a^4 \omega^2}{4g} = \dfrac{\pi a^2 b}{2}$. $\therefore \omega = \dfrac{1}{a} \sqrt{2gb}$.

Ex. 3. A cylinder is 2 feet in diameter and 4 feet deep, and is whirled round its vertical axis till a point in the circumference passes through 40 feet per second. If the cylinder were originally full of water, find the number of cubic feet thrown out. Ans. $11\frac{5}{9}$.

CHAPTER XI.

ON GIRDER BEAMS AND BRIDGES, THE STRENGTH OF TUBES, AND THE CATENARY.

175. We have already explained the meaning of the word *elasticity*, as indicating that property of matter whereby all substances tend to recover their original dimensions when the forces to which they are subjected return also to their original intensity. At present we refer to solid bodies.

LAWS OF ELASTICITY.

Take the case of a slender rod, say of wrought iron, some 3 or 4 feet long, suspended from one end, and stretched by weights hung from the other end.

1. *The amount of extension for different weights will be in direct proportion to the stretching weight.*

This law is verified very approximately by experiment.

2. *The amount of extension is in direct proportion to the length of the rod.*

It is evident that this law must hold, for the extension in every part of the rod is the same. Thus, if a spiral steel spring, one foot long, be stretched one inch under a pull of

10 lbs., a similar spring two feet long will be stretched two inches under the same pull of 10 lbs.

3. *The amount of extension is in inverse proportion to the sectional area of the rod.*

That is to say, on doubling the area of the rod the resistance will be doubled, and the extension will be diminished one-half.

4. *The amount of extension for rods of different material depends on the nature of the material.*

Hence we assign to every different substance a certain numerical co-efficient called the *modulus of elasticity*, which enables us to say beforehand how much a rod of a given form and material will elongate under a given strain. This *modulus* represents an imaginary fact, which could only be realised in such substances as india-rubber, viz., that a weight in pounds, given by the number registered as the modulus, would stretch a square inch bar of the given material to double its length. The *modulus* is an enormous number, and tables of its value are given in technical treatises. For wrought iron plates it is 29,000,000 lbs. ; for copper wire, 17,000,000 lbs. ; for English oak, 1,450,000 lbs.; for sheet lead, 720,000 lbs.

In order to express these laws by a formula, let l be the length of a rod before straining it, A its sectional area, P the stretching weight, c the modulus of elasticity. Then the elongation produced by $P = \dfrac{P l}{A C}$.

If L be the length of the rod after elongation, we have

$$L = l + \frac{P l}{A C}.$$

Note. These remarks apply equally when the force tends to compress the rod, the fundamental law being that the amount of compression is in direct proportion to the compressing force. In order to avoid confusion we have spoken only of elongation, but, *mutatis mutandis*, we deal in the same manner with compression.

Ex. Find the extension of a bar of wrought iron of $\frac{1}{2}$ square inch sectional area, the bar being 4 feet long, and loaded with 4 tons.

Here 4 tons = 8960 lbs., $\therefore$ extension = $\dfrac{8960 \times 48}{\frac{1}{2} \times 29,000,000}$ = ·03 inch.

LIMITS OF ELASTICITY.

176. It is convenient to examine a law by means of a formula, and the expression just found enables us to form a very clear idea of what is meant by the *limits of elasticity*. So long as the law holds the extension is in some fixed proportion to P; and when P = 0, L becomes equal to *l*. Every one must have observed that it is possible to bend or stretch rods of metal out of shape without breaking them, hence we say that every substance has a *limit of elasticity*, and this limit is known when we assign the greatest value of P which does not produce any permanent deformation in the specimen. When P is removed, the length L should again become equal to *l*, as at first. The limit of elasticity of wrought iron is about 10 tons on the square inch, and each ton of strain will lengthen a bar $\frac{1}{10,000}$th part.

Note. In practice the term *limit of elasticity* is usually considered with reference to stretching forces, but there is a limit of elasticity for compression, just as much as for extension, the test being that the body perfectly recovers its original dimensions when the compressing force is removed.

Note. As soon as we have passed the limits of elasticity we observe a new property, viz. *ductility*, which it is essential to register. Let specimens of iron, and steel be prepared, 2 inches long in the central part, which is cylindrical, and ·5 of an inch in sectional area. Place the specimen in a testing machine until it is pulled asunder into two pieces. It will elongate and yield throughout, but most of all near the centre, where it will finally break. By fitting the two parts together and carefully measuring the length of the cylindrical portion, we can ascertain how much it has elongated. If the cylindrical part be 2 inches in length before testing, and 2$\frac{1}{2}$ inches long after it is broken, we should call the ductility 25 per cent. It must be understood that the only part of the specimen which can yield is the cylindrical portion. Sir J. Whitworth states that the tensile strength, or breaking weight, for a square inch bar of Lowmoor iron is about 27 tons, with a ductility of 38 per cent. The tensile strength of one kind of fluid compressed steel has been shown to be 40 tons, with a ductility of 32 per cent.; and when the strengths increase to 48, 58, 68 tons per square inch, the ductilities will fall to 24, 17, and 10 per cent. The tensile strength of good cast iron is 10 tons, with a ductility of $\frac{3}{4}$ per cent.

WORK DONE DURING EXTENSION.

177. *Prop.* A bar whose length is l is stretched to a length $l + c$ by a force P lbs.; to find the work done during the extension, which lies within the limits of elasticity.

This is an example of work done against a varying resistance, which increases in a constant ratio, for if the bar stretches through a length x under a pull of 1 lb., it will stretch through $2x$ under a strain of 2 lbs., and so on. Take two straight lines A B, B C at right angles to each other; let B C represent P, and let A B represent c. Join A C, take p, q, two points near together in A C, and draw $p\,m, q\,n$

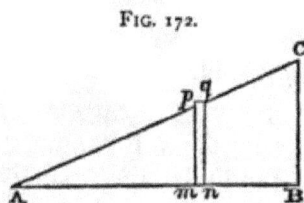

FIG. 172.

perpendicular to A B. It is clear that A m represents the elongation produced by the force $p\,m$. Conceive that the force equal to $q\,n$ effects the next addition of elongation, viz., $m\,n$, and is constant throughout, then the work done by $q\,n$ will be represented by the rectangle $q\,m$. Thus the whole work done in stretching the bar through A B is represented by the triangle A B C, or by the expression $\dfrac{P\,c}{2}$.

Ex. A force of 10 lbs. stretches a spiral spring through 1 inch; what work is done in extending the spring through $\frac{3}{4}$ inch?

Here A B = 1 inch, and B C represents 10 lbs., therefore the work

required $= \frac{1}{2} \times \frac{3}{48} \times \frac{30}{4}$ foot pounds $= \frac{15}{64}$ of a foot pound.

THE TRANSVERSE STRAIN UPON BEAMS.

178. It has been explained in *Art.* 32 that a *transverse strain* upon a beam is produced by a force acting perpendicularly to the direction of the beam. We purpose now to calculate the dimensions of rectangular beams of any selected material which shall be capable of sustaining given amounts of transverse strain. The problem is divided into two parts :

1. We calculate the *moment* of the forces arising from the weight of the beam, or of the load upon it, and we estimate that moment about any point which we may choose to select. This gives the moment of the breaking force about any point of the beam.

2. We calculate the resistance which the fibres of the beam offer to this fracture, and we proceed on the hypothesis that the fibres of the beam constitute an assemblage of elastic bars, which are capable of resisting both extension and compression. In this way we obtain the sum of the moments of the resistance of the fibres about those points in the beam which are neither extended nor compressed.

By the principle of the lever we equate the moment of the breaking force to the sum of the moments of the resisting fibres, and deduce the law of strength of the beam.

179. *Prop.* To find the moment of the strain which the fibres of a rectangular beam exert in resisting extension and compression at the breaking point.

Let D A B C represent a section of the beam by a plane perpendicular to its axis, D A $= b$, D C $= d$, and conceive that the beam is supported on the horizontal line C B, while loaded at one end by the weight w. The beam itself is drawn on a somewhat reduced scale.

We confine our attention to a narrow vertical strip of fibres along A B, and whatever is proved for that strip will be equally true for every other similar strip till we arrive at D C.

It is evident that the fibres along A B will be extended on the side A, and compressed on the side B, and that somewhere

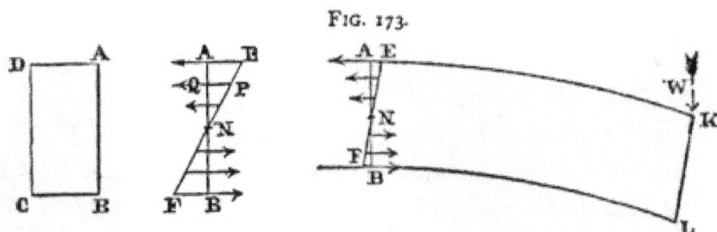

between A and B will be a point N, where the fibres are neither extended nor compressed. The position of N is taken to be in the centre of gravity of the mass of fibres, in this case the centre. We have shown the fibres in F L as being compressed through B F, while those in A K are extended through A E; the arrows indicating the action of some of the fibres in pulling against W where extended, and in pushing against it where compressed.

Let $A E = B F = e$, $Q N = x$, and let the force producing the extension P Q on a set of fibres at Q, whose sectional area is one square inch, be m P Q, where m is some constant. Hence the tension on a slice of fibres at Q, of breadth b and depth $dx = m$ P Q $\times b \, dx$.

Now $PQ : e = x : AN = x : \dfrac{d}{2}$, $\therefore PQ = \dfrac{2 e x}{d}$,

substituting this value for P Q, and taking moments about N, we have the moment of the tension on the slice at Q

$$= \frac{2 \, m \, e \, b \, x^2 \, d x}{d}.$$

Hence the sum of the moments of the tension of the fibres above N $= \dfrac{2 \, m \, e b}{d}$ (sum of $x^2 \, d x$).

It may be proved by analysis that the sum of a series of quantities such as $x^2 \, dx$, taken from $x = 0$, to $x = \frac{1}{2} d$ is $\frac{1}{3} \dfrac{d^3}{8}$, therefore the sum of the moments of the tension of the fibres above N $= \dfrac{m \, e b \, d^2}{12}$.

The same is true of the sum of the moments of compression of the fibres below N; and therefore the whole sum of the moments, both of extension and compression,

$$= \frac{m\,e\,b\,d^2}{12} + \frac{m\,e\,b\,d^2}{12} = \frac{m\,e\,b\,d^2}{6} = s\,b\,d^2,$$

where s is a number to be determined by experiment.

Note. If N be not in the centre of A B, we shall still have the sum of the moments of the extended and compressed fibres respectively varying as $b\,d^2$, and we may take their sum as $s\,b\,d^2$; whence we observe that in all cases the moment $s\,b\,d^2$ is to be made equal to the moment of the strain of the external forces. The coefficient s is registered in tables, or can be found by experiment. In the following examples the weight of the beam is neglected :—

Ex. 1. A batten of Riga fir, 7 feet long, 2 inches square, and supported at its extremities, is broken by a weight of 422 lbs. suspended in the centre. Find s.

Here we take the formula in Ex. 2, page 100, therefore

$$\frac{w\,l}{4} = s\,b\,d^2, \text{ or } \frac{422 \times 7 \times 12}{4} = s \times 2 \times 4, \therefore s = 1108.$$

Ex. 2. A bar of cast iron is 3 feet long and 1 inch square, and it is broken by a weight of 844 lbs. Find s.

Here $844 \times 9 = s \times 1, \therefore s = 7596.$

Ex. 3. In a beam of Riga fir, $l = 5$ feet, $b = 4$ inches, $d = 6$ inches, the beam being supported at one end and loaded at the other end; find the greatest weight that the beam will support.

Here $w\,l = s\,b\,d^2, \therefore w = \dfrac{24 \times 1108}{10} = 2659 \cdot 2$ lbs.

Ex. 4. A beam of Riga fir, 20 feet long and 12 inches square, is supported at both ends; find the weight which it will support at a distance of 8 feet from one extremity.

Here $s\,b\,d^2 = w \times \dfrac{96 \times 144}{20 \times 12}$,

$\therefore \quad \times 1108 \times 12 \times 144 = w \times 96 \times 12, \therefore w = 33240$ lbs.

Ex. 5. A cistern which holds a ton of water is supported on two beams of larch (s = 1127) projecting 4 feet from a wall. Each beam is 5 inches deep, what should be its breadth?

ROLLED IRON BARS.

180. Having proved that the strength of a rectangular beam varies as (breadth) × (depth)^2, we must endeavour to dispose the material of an iron beam in a more advantageous manner than is practicable with timber. There are three common forms of rolled bar iron used in construction, known as *angle iron*, *T iron*, and *channel iron*, where *the strength due to increased depth* is exhibited.

FIG. 174.

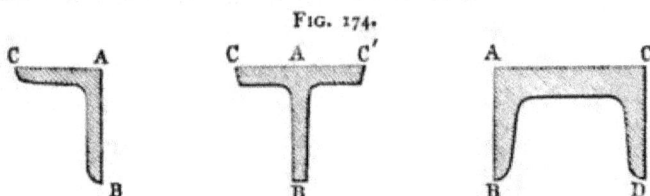

1. The *angle iron* has two elongated sides A C, A B giving it strength in two perpendicular directions. It is in fact a compound beam.

2. The *T iron* is a flat weak beam C A C′, strengthened by a vertical flange A B.

3. The *channel* or *double angle iron* is an improved form of T iron, relying for its strength on the two vertical flanges A B, C D. It is extensively used in the construction of iron bridges, and, as often happens in mechanics, it has proved useful in very homely applications. The frames of umbrellas are now made of steel wire rolled in the form of channel iron, and are stronger and lighter than those of the old-fashioned construction, where the ribs were solid square wires.

CAST IRON BEAMS.

181. In the previous examples the dimensions are commonly small, the length of A C in the channel iron, as sketched, being 3 or 4 inches; and we pass on to the construction of a beam of *cast iron*, used for supporting a roadway, say 20 to 30 feet in length, and technically called a *girder beam*. The object now will be to place the material where it can act most effectively, and that is as far as possible from N.

In the case of the rectangular beam the fibres at A and B are acted on mroe powerfully than thóse nearer to N, and as soon as they yield the beam will break. If we could condense the material into two parts about A and B, merely connecting them by a web strong enough to prevent the beam from bending out of shape or *buckling*, we should obtain greatly increased strength from the same weight of metal. Mr. Hodgkinson has investigated the dimensions of the best form of cast iron beam, and we give a sketch of the beam so determined.

FIG. 175.

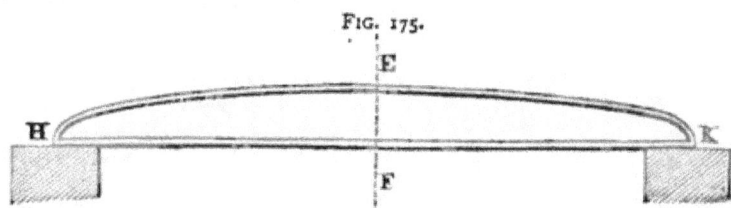

The resistance of cast iron to direct crushing is more than 6 times its resistance to tearing. Hence the area of the top flange A B, which resists compression, may be $\frac{1}{6}$ that of the bottom flange C D, which resists extension. In practice the area of the top flange is often $\frac{1}{4}$ that of the lower flange. The drawing shows a transverse section of the beam at E F.

FIG. 176.

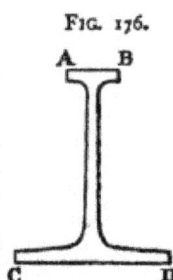

The moment of the breaking strain due to a load uniformly distributed will increase as we approach the centre of the beam, hence the distance between the flanges A B, C D is increased from H or K to F, and the beam has the curved outline given in the sketch. There is not space to discuss this matter fully, but it may be an advantage to consider the law which governs the form of a beam. Let K C A represent a beam of *uniform* breadth whose section is A B C D, and conceive that the beam is supported at the enlarged

T

end and loaded at к with a weight w. Our object will
be to find the form of the curve к р с when the beam

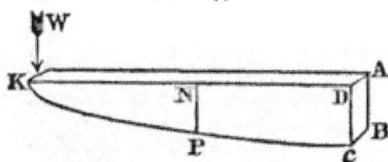

FIG. 177.

is equally strong throughout.
Draw p n vertical and let
к N $= x$, N P $= y$, A D $= b$,
then the moment of the
strain at N $=$ w x, while the
moment of the resistance of
the fibres $=$ s (breadth) (depth)$^2 = s\,b\,y^2$.

$$\therefore w\,x = s\,b\,y^2, \text{ or } y^2 = \frac{w}{s\,b}\,x.$$

which is the equation to a parabola, and therefore the curve
к р с is a parabola. The student will have no difficulty in
connecting this proposition with the curved outline of the
beam adopted in a steam-engine.

WROUGHT IRON BEAMS.

182. The resistance of wrought iron to extension is somewhat greater than its resistance to compression, and may be
taken at 4 tons per square inch in compression, or 5 tons
per square inch for extension. Hence the upper and lower
flanges in a wrought iron beam are nearly, and often
quite, equal in area. Also the material can be used for
beams of great length, and we take the girder beams in the
Cannon Street Bridge as the types of this form.

1. The *Box-girder*, in which the upper and lower members

FIG. 1 .

A B, C D, are formed of a series of
plates rather more than $\frac{1}{2}$ inch thick,
connected by plate webs A C, B D.
For this particular bridge they are
made 8 feet 6 inches high, 3 feet
7 inches wide, and 125 feet span.

2. The *Plate-girder*, in which the
upper and lower members A B, CD are
formed of a series of plates connected
by an intermediate web. The span
and depth are the same as before,

but the width is 2 feet 2 inches. The plate and box girders are strengthened by angle or T irons, as may be necessary; there are also internal diaphragm plates in a large box-girder. The roadway lies on the top of the girders.

Note 1. In these wrought iron beams the increased strength required as we approach the centre is obtained without departing from the straight line form by putting additional plates on the top and bottom flanges.

Note 2. In the Britannia bridge over the Menai Straits the beam is a hollow rectangular iron tube, 472 feet in length, and the parts A B, C D, are formed of hollow beams somewhat in the proportion indicated in the sketch. There are two main tubes, each of which weighs 5,270 tons. The necessary strength is obtained by the assemblage of hollow beams lying along the top and bottom of the bridge. The height of the tube is 28 feet, its breadth is 14 feet; there are 8 cells $1\frac{3}{4}$ feet square along A B, and 6 cells $2\frac{1}{3}$ feet square along C D.

FIG. 179.

THE PRINCIPLE OF THE WARREN GIRDER.

183. The ordinary plate girder is the best form for moderate lengths, but is not adapted for very large spans. In both the previous types of girder beam the strength is distributed along the whole of the two lines A B, C D, and it will manifestly be a better arrangement to concentrate it in the four points A, B, C, D. The beam would then become compound, and would consist of two parallel girders, one occupying each vertical side of the rectangle A B C D. A straight-line bridge would thus consist of two outside girder beams, the roadway running between them, and each girder beam A C, B D, would be formed of two distinct members, to be connected in some way so as to make a perfect beam. The first important example of this type of bridge was the Newark Dyke Bridge, on the Great Northern Railway, which had a span of 240 feet. Here the *upper member* consists of a number of cast iron tubes, increasing somewhat in size and thickness towards the centre, and rather more than a foot in diameter. These tubes support the compression of the upper portion, and are placed at the angles A, B. The *lower member* is formed of a series of wrought iron links, each 18½ feet long, 9 inches deep, and from 1 to 1½ inches wide. These links are placed edgeways, and lie side by side at the angles C and D, their number being increased from 4 at the piers to 6, 8, 10, 12, and finally to 14 at the centre of the bridge. In this way the resisting power of the beam is increased in proportion to the moment of strain.

The next point to be studied is the method of connecting the upper and lower members of either beam. This is effected, not by a continuous web, but by a series of equilateral triangles, presenting one type of *lattice-girder*. In order to understand the mechanical action which has to be met when the beam is loaded, conceive that two laths of wood are laid upon one another, fastened at the centre, and supported at the two ends. If this compound beam be loaded in the middle, it will become bowed, and the ends of the upper strip will overlap the ends of the lower strip. In other words, each half of the upper member tends to move from the centre outwards, when the beam is under the strain of a load. The object of the triangles is to prevent this motion, for the beam cannot bend to any serious extent when such a tendency is overcome.

FIG. 180.

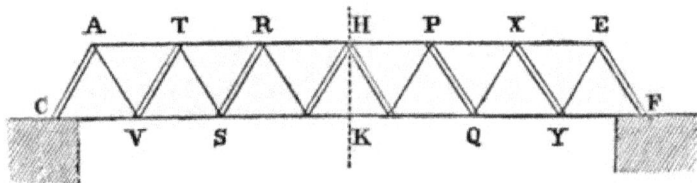

Let A E, C F represent the upper and lower members of a Warren girder, H K the central line. Our proposition is that when the beam is loaded H A tends to slide a little from H outwards relatively to K C, as does also H E relatively to K F. It is evident that a series of struts* A C, T V, R S, will keep H A in position relatively to K C and a series of ties A V, T S, will assist the action. The struts should all incline towards the centre of the beam. In like manner H E will be kept in its place relatively to K F by a series of *struts* P Q, X Y, E F, all pointing towards H, and a series of *ties* X Q, E Y acting to pull against any movement in H E.

This matter may be made very clear by experiment. Let two bars of wood A E, C F, be connected by a series of metal strips, such as R S, X Q, all pointing in one direction.

FIG. 181.

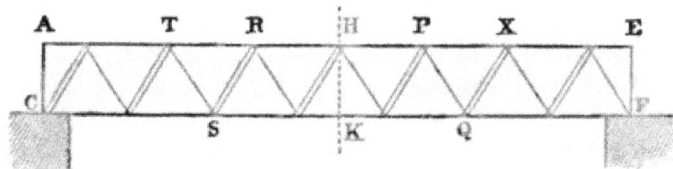

and also by a series of strings such as T S, P Q, it will be found that the half beam C H is perfectly strong, but that H F will bend out of shape directly. The construction is mechanically right from C to K, and wrong from K to F. The ties are placed where the struts ought to be, and whereas the strings from A to K will tighten under the load, those from K to F will altogether lose their tension.

The same thing may be shown with a model such as that represented in Fig. 180. Let the beam be supported at C and F, and loaded in the centre, the strings will all tighten, the struts will act, and the structure will remain rigid and immovable in all its parts. But turn the same beam over, and let it rest on the ends A, E. Every string will become slack, the beam will bend out of shape, and may be broken quite easily.

---

* A *strut* is a bar which resists compression, a *tie* resists extension.

The last point to be noticed is, that the struts and ties are made lighter and less massive as we approach the centre of the bridge. This is the opposite of what happens with the upper and lower members, and would not be anticipated without careful reasoning. Conceive that equal weights are hung at the vertices of each triangle in the model. Any triangle will form an isosceles roof with a tie across the base, and there will spring from each vertex a pair of equal and opposite horizontal forces travelling right and left along the upper member. Fixing our attention on the vertex of any particular triangle, we observe that a group of forces pointing to the left will come to it from every vertex on the right hand, and a group of forces pointing to the right will come to it from every vertex on the left hand. These will partially counterbalance, and the surplus will be greatest at each pier and zero at the centre. Hence the struts and ties must be stronger as we approach the piers. This peculiarity referred to is very apparent in the railway bridge at Charing Cross. This bridge also illustrates the use made of channel iron, the upper and lower lines of the girder being constructed of plates in the form of channel iron with four vertical ribs.

In order to give an idea of the strength of one of these straight line bridges, we may state that the Newark bridge was tested by distributing 240 tons upon the entire length from pier to pier, when the deflection at the centre amounted to $2\frac{3}{4}$ inches. The Warren girder has been described on account of its simplicity and because it has suggested better forms. The steps from a Warren to a lattice girder are exemplified in the Charing Cross and Blackfriars bridges.

THE STRENGTH OF CYLINDERS UNDER INTERNAL PRESSURE.

184. There is scarcely space to touch briefly on another subject, viz., the estimation of the thickness of cylindrical steam boilers, or of pipes conveying water under pressure.

Prop. To find the thickness of a cylindrical boiler required to support a given pressure of steam.

Let e be the required thickness of the cylinder, r the radius of its inner surface, w the tensile strain per square inch which the material of the tube is estimated to support. We shall consider the action upon a small portion of the tube shown in the sketch whose sides are x, y, and whose thickness is e.

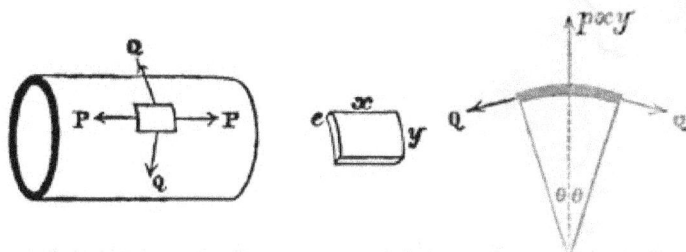

Let p be the pressure of the fluid, then $p\,x\,y$ is the pressure on the area $x\,y$. Also $e\,x$ is the area of the section which supports a pressure Q. Now 1 square inch support w lbs., $\therefore e\,x$ square inches will support w $e\,x$ lbs., or $Q = w\,e\,x$.

But by the resolution of forces we have $2\,Q \sin \theta = p\,x\,y$,

and $\sin \theta = \theta$ (very nearly) $= \dfrac{y}{2r}$,

therefore $2\,w\,e\,x \times \dfrac{y}{2\,r} = p\,x\,y$, $\therefore e = \dfrac{p\,r}{w}$.

Note. Substituting for e we obtain $Q = p\,r\,x$.

Ex. A cylindrical boiler is 4 feet in diameter, and is required to support a pressure of 100 lbs. on the square inch. The tenacity of riveted plate iron being 34,000 lbs. per square inch, what should be the thickness of the material?

Here $e = \dfrac{p\,r}{w} = \dfrac{100 \times 24}{34,000} = \dfrac{3}{14}$ inch nearly.

If the material is not to support a greater strain than 5,000 lbs. on the inch, we have $e = \cdot 5$ inch nearly.

185. *Prop.* To find the thickness of the cylinder necessary to prevent the flat end from being torn off.

The pressure on the flat end is $p \times \pi r^2$, and this pressure is supported by the cohesive strength of the material in a transverse section whose area is $2 \pi r e$. As before, $2 \pi r e$ square inches will support $2 \pi r e w$ lbs. Hence

$$2 \pi r e w = p \pi r^2, \text{ and } e = \frac{p r}{2 w},$$

which is exactly half the value formerly obtained.

To make this more intelligible, we shall find P and contrast it with Q. Now P is the pressure on an area $e y$, being part of a ring whose area is $2 \pi r e$. Therefore

$$\frac{P}{\pi r^2 p} = \frac{e y}{2 \pi r e}, \text{ or } P = \frac{p r y}{2}.$$

Collating the results we have $Q = p r x$, $2 P = p r y$. Take $x = y$ in our small section, then $Q = 2 P$, or the longitudinal strain is twice the transverse strain. Hence a tube under pressure is most likely to yield and split open in direction of its length, and it may be strengthened very materially by means of rings which embrace it at intervals. Thus we sometimes see rings cast round very long cylinders in a steam engine, and we can understand the great increase of strength which a pipe receives from the flange. The rule that the strength of a tube is inversely proportional to the diameter is well known, and pipes for conveying water under pressure are always comparatively small in size. There is another rule also which we can deduce without calculation. Setting aside any weakness due to rivets, *the hemispherical end of a cylindrical boiler is twice as strong as the cylindrical portion.* If the section $x y e$ were spherical we should have the forces P, P inclining inwards as well as Q, Q. Hence we should have a doubling of the force which mechanically pulls against the outward bursting pressure, and deduction is manifest.

Sir Joseph Whitworth has made some interesting experiments on the strength of fluid-compressed steel by the *gunpowder test.* Cylinders of the metal are turned and bored, and are 4 inches long, $1\frac{1}{4}$ inches external diameter, $\frac{3}{4}$ inch internal diameter. A small charge of powder is inserted in the centre of the tube, which is then plugged with wax, and closed at the ends by copper cups, similar to the cup leathers

FIG. 183.

A   B   C

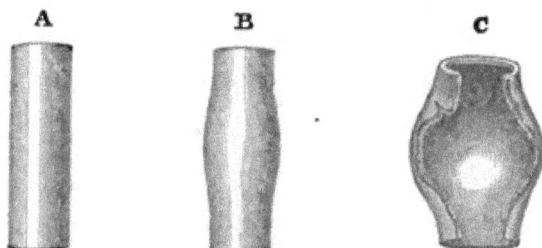

of a press. The tube is held in position by screwed plugs inserted into a massive cylinder. Successive powder charges are fired, which are increased as the tube yields to pressure and bulges at the centre. The enlargements for each successive discharge are registered, and finally the tube bursts by tearing along one longitudinal line, as shown in the drawing, which also gives the tube before firing and after several charges have been fired. A good specimen will not break into pieces, and our theory has shown that it may be expected to yield along a longitudinal section. When the material is thickened it becomes absolute master of the gunpowder. Thus a cylinder of fluid-compressed steel 2·56 inches internal diameter, and 2·63 inches thick has supported the strain of a charge of $1\frac{1}{2}$ lbs of powder more than fifty times without any enlargement worthy of notice. The only escape for the powder gas has been through a touch hole $\frac{1}{10}$ inch in diameter, and after forty-eight discharges the enlargement of the outside diameter was ·0485 inch.

THE CATENARY CURVE.

186. When a chain is hung from two distant points, the curve which it assumes is called a *catenary*.

Let E A D represent a chain hung from two points E and D

in the same horizontal line, A being the lowest point of the curve. We shall adopt the artifice of supposing the chain to become rigid, and can then apply the conditions of equilibrium of forces acting on a body in one plane.

Note. The chain is of uniform thickness.

Let $A P = s$, and assume that w is the weight of an unit of length of the chain, then $w s$ is the weight of A P. Also let $w a$ and $w t$ represent the tensions at A and P. Then

FIG. 184.

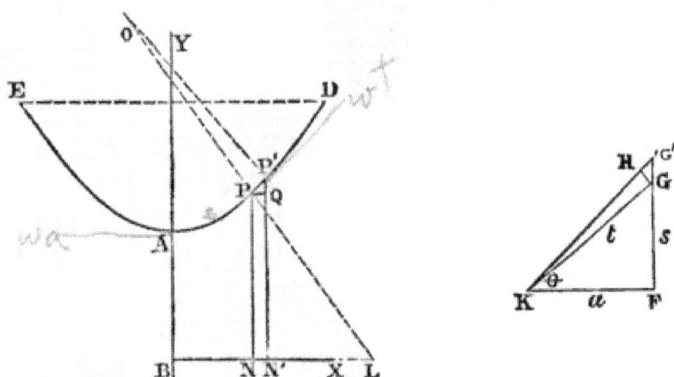

the portion A P is at rest under (1) the pull at A, which is a horizontal force, (2) the pull at P, which acts in the tangent at P, (3) its weight, which acts vertically. Construct a triangle K F G whose sides are parallel to these forces, viz. K F parallel to $w a$, K G parallel to $w t$, G F parallel to $w s$.

Then, by the triangle of forces, if

K F $= a$, we shall have K G $= t$, and G F $= s$.

Next draw A B vertical and equal to a, take B A Y, B X as rectangular axes to which the curve is to be referred, and let $B N = x$, $P N = y$.

1. *To prove that $y = t$.*

Let P P′ be a small arc of the curve, draw K G′ parallel to the tangent at P′, and G H perpendicular to K G′. In like manner draw P′N′ vertical and P Q perpendicular to P′N′. Let $A P′ = s′$, and conceive the arc P P′ to be so small that its curvature may be neglected, then the triangle P Q P′ is equal

in all respects to G H G'. Also let accented letters refer to
P' instead of P, then $GG' = s' - s$, $GH = PQ = x' - x$,
$P'Q = y' - y$, $HG' = t' - t$,

but $P'Q = HG'$, $\therefore y' - y = t' - t$.

Hence the increase of y is equal to the increase of t, but
y and t are equal at the point A, therefore they are always
equal, or
$$y = t.$$

2. *To find the equation to the curve.*

Since the triangle G H G' is similar to K F G, we have

$$\frac{GG'}{KG} = \frac{G'H}{GF} = \frac{GH}{KF}, \quad \therefore \frac{GG' + G'H}{KG + GF} = \frac{GH}{KF}$$

or $\quad \dfrac{s' + y' - (s+y)}{s + y} = \dfrac{x' - x}{a}$.

Hence by a theorem in algebra (see *Art.* 67) we have

$$\log(s' + y') - \log(s + y) = \frac{x' - x}{a}.$$

That is, the increase of log. $(s + y)$, in passing from P to P',
is equal to the increase of x divided by a, and this being
true for every point is true on passing from A to P.

$$\therefore \log(s + y) - \log(o + a) = \frac{x - o}{a} = \frac{x}{a},$$

or $\log \left(\dfrac{s+y}{a} \right) = \dfrac{x}{a}$, and $s + y = a \, e^{\frac{x}{a}}$, \hfill (1)

similarly $y - s = a \, e^{-\frac{x}{a}}$, \hfill (2)

therefore $2y = a \left(e^{\frac{x}{a}} + e^{-\frac{x}{a}} \right)$,

which is the equation to the catenary curve.

Cor. Equations (1) and (2) give $2s = a \left(e^{\frac{x}{a}} - e^{-\frac{x}{a}} \right)$.

3. *To prove that the radius of curvature at any point of the
catenary is equal to* $\dfrac{y^2}{a}$.

Draw normals at P and P' intersecting in o, the centre of
the circle of curvature, and let $OP = R$, $POP' = \alpha$.

Then $PP' = R\alpha$, $GG' = PP' = R\alpha$, and $HG = t\alpha$,

but $\dfrac{GG'}{HG} = \dfrac{KG}{KF} = \dfrac{t}{a} = \dfrac{y}{a}$,

$$\therefore \frac{R\,a}{t\,a} = \frac{y}{a}, \text{ or } \frac{R}{y} = \frac{y}{a}, \text{ or } R = \frac{y^2}{a}.$$

Cor. 1. The normal P L, obtained by producing O P to meet B X in L, is also equal to R.

$$\text{For } \frac{PL}{y} = \frac{KG}{KF} = \frac{y}{a}, \therefore PL = \frac{y^2}{a} = R.$$

Cor. 2. The catenary has the property that its centre of gravity is further below the horizontal line E D than it would be if the chain were to assume any other arbitrary form. This is evident, for each portion tends to get as low down as possible, and therefore the centre of gravity does the same. In any other arbitrary curve some portion of the chain would be raised and the centre of gravity would also be raised.

Ex. 1. Let E D = 800 feet, and suppose that a = 1,600 feet, find the length of the chain, and the depth of A below E D.

Here $\frac{x}{a} = \frac{1}{4}$, and the depth of A $= y - a = \frac{a}{2}(e^{\frac{1}{4}} + e^{-\frac{1}{4}}) - a$.

Now $e = 2\cdot71828$, $\therefore e^{\frac{1}{4}} = 1\cdot284, e^{-\frac{1}{4}} = \cdot7788,$

$$\therefore \text{ depth of A} = 50\cdot24.$$

Also A D $= \frac{a}{2}(e^{\frac{1}{4}} - e^{-\frac{1}{4}}) = 800 \times \cdot5052 = 404\cdot16.$

We can also find the inclination of the curve to the horizontal line D E at either point of suspension. Let θ be this angle, then

$$\cos\theta = \frac{KF}{KG} = \frac{1600}{1650\cdot24}, \therefore \theta = 75°\,46'.$$

Ex. 2. A chain 110 feet long is suspended from two points in the same horizontal plane at a distance of 108 feet ; find the tension at the lowest point.

In this example the chain assumes a nearly circular form, and the radius of curvature at the lowest point is $\frac{a^2}{a}$ or a.

FIG. 185.

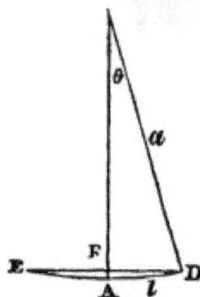

Let A D $= l$, and we have $\frac{FD}{a} = \sin\theta = \sin\frac{l}{a}$,

But $\sin\theta = \theta - \frac{\theta^3}{6}$, when θ is small,

$$\therefore \frac{FD}{a} = \frac{l}{a} - \frac{l^3}{6a^3}, \text{ or } a = \frac{l^3}{6(l - FD)} = \frac{55^3}{},$$

$$\therefore a = 166.$$

In order to verify the result we will reverse the problem, and find s when $a = 166$.

Here $e^{\frac{54}{166}} - e^{-\frac{54}{166}} = 1\cdot3844 - \cdot72233 = \cdot6621.$

$\therefore s = 83 \times \cdot6621 = 54\cdot95$, whence $E A D = 109\cdot9$.

187. If the thickness of the chain varies it may hang in curves of varied form, and our present object is to ascertain the law of distribution of weight, in order that the chain may assume any required curvature.

Adopting the previous notation, let $P P' = s'$, and let w' be the weight of $P P'$. If $F G$, in Fig. 184, represents the weight of $A P$, we may take $G G'$ to represent w', also $s' = R a$, and $a = \dfrac{H G}{K G}$, therefore

$$\frac{s'}{R} = a = \frac{H G}{K G} = \frac{G G' \times F K}{K G^2} = \frac{w' a}{t^2},$$

$$\therefore \frac{w'}{s'} = \frac{t^2}{a R} = \frac{a}{R} \times \frac{t^2}{a^2} = \frac{a}{R} \sec^2\theta.$$

Ex. 1. Let the curve be a circle of radius c,

$$\therefore \frac{w'}{s'} = \frac{a}{c} \sec^2\theta.$$

Ex. 2. Let the curve be a parabola.

Here $P G = n$, $N G = l$, $P G$ being the normal at P,

$$\therefore \frac{w'}{s'} = \frac{a}{R} \sec^2\theta = \frac{a}{R} \times \frac{n^2}{l^2}.$$

But $R = \dfrac{n^3}{l^2}$, by a property of the parabola.

FIG. 186.

$$\therefore \frac{w'}{s'} = \frac{a l^2}{n^3} \times \frac{n^2}{l^2} = \frac{a}{n},$$

or the thickness varies inversely as $P G$.

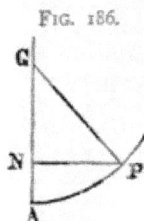

THE PRINCIPLE OF THE ARCH.

188. An inverted catenary would form an arch, and we have introduced the investigation into the properties of the catenary curve in order to lead the student to form the first simple conception of the mechanical principle of the arch. In considering the equilibrium of a flexible chain hanging from the points E D, it is now evident that any portion such as $A P$, will be at rest under (1) the pull at P, (2) the pull

at A, (3) the weight of A P. If the chain were to become rigid and inflexible the forces acting would not be changed, and if the rigid curve were inverted we should obtain a linear

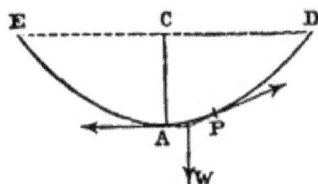

Fig. 187.

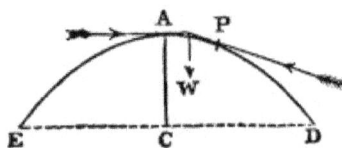

Fig. 188.

arch which would be at rest under the action of the like forces, with the single exception that the pulls at A and P would now be converted into thrusts supporting A P from below. What we should call a line of tension in the case of the ordinary catenary would become a line of pressure, and the curve itself would indicate the direction in which the force at every point of it was exerting its action.

189. An arch is an assemblage of wedge-shaped masses, taking the form of a ring, and supported by their mutual pressures. A portion of an arch is shown in the drawing. The separate stones are called *voussoirs*, the top stone is the *key stone*, the surfaces *c d*, *e f*, between the stones are *joints*, the internal curve is the *intrados*, the external curve is the *extrados*, and the supports are *piers* or *abutments*.

There is a theoretical possibility of equilibrium when the

Fig. 189.

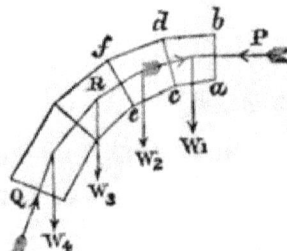

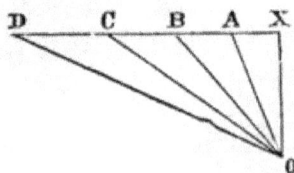

joints are smooth. Thus the voussoir *a b d c* is at rest under its weight, and the pressures in directions of the arrows. We suppose *a b* to be vertical and draw o x vertical. The line

O X may be taken to represent the pressure on *a b*, while X A perpendicular to O X may represent the weight of *a b d c*, and O A drawn parallel to *c d* will then represent the pressure on C D. Proceeding in this manner, we may draw O B, O C .. parallel to the respective bed-joints, and A B, B C, . . . will represent the respective weights of the voussoirs. Also the line Q R P will be the line of pressures.

From what has preceded, we infer that if the line of pressures of an arch ring were a flexible string loaded at proper intervals with the weights of the voussoirs, it would not alter its form when suspended at the two ends; in other words, the line of pressures always preserves its relationship to a weighted catenary curve. Again, if the joints be smooth, the weight of each voussoir must remain a definite quantity, and cannot be varied; hence the theoretical arch will not support the smallest extra load on any part of it. Whereas, if the joints be rough, the weight placed on any voussoir may be increased until the angle which the line of pressures makes with the perpendicular to the corresponding joint is equal to *the angle of repose* for the surfaces in contact, and this angle may be made as large as we please by roughening or *joggling* the stones. This shows the value of friction in preventing the voussoirs from slipping, and thereby enabling an arch to stand when supporting a load.

But the arch, when heavily loaded, is liable to a more

FIG. 190.

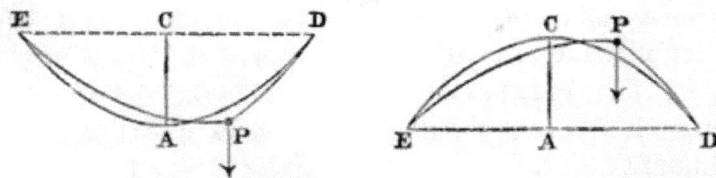

imminent danger than the slipping of the voussoirs, for the line of pressures may come to the extreme end of a joint, in which case there is nothing to prevent the arch from opening, and the structure yields in the manner about to be described. In order to trace the action, let us examine the effect of a load upon the line of pressures. For this purpose, hang a weight P at some point of the catenary curve, and the line of tensions will take the new form shown in the diagram. If the chain became rigid and were inverted, we should have a line of pressures with a raised apex, the two portions of which point more directly towards the weight they are called upon to support.

Conceive now that an arch ring is unduly loaded with a weight W, the line of pressures will rise up towards the weight, and it may happen that in doing so the line is shifted to the extremities of the joints at A, W, C, D. The

FIG. 191.

arch may split up into three portions, and may open at each of the joints A, B, C, D. The separate pieces are in the same condition as a body resting on a plane, and overhanging to such an extent that the vertical through the centre of gravity falls on the edge of the base. The portion B will come lower, but C will rise, and the arch will fall by the opening of its joints. Thus the conditions of equilibrium of an arch become much more intelligible when we understand the effect of a load in changing the direction of the line of pressure. The safeguard consists in preserving the line of pressures well within the limits of the arch ring under all conditions of load. Also since W cannot descend unless C also rises, we comprehend that an arch built in a wall is almost sure to stand.

CHAPTER XII.

ON SOME MECHANICAL INVENTIONS.

IN this concluding chapter we shall examine a few well-known inventions, which exhibit the application of mechanical principles.

THE PRINCIPLE OF GIFFARD'S INJECTOR.

190. This invention furnishes an illustration of the direct conversion of heat into mechanical work, and opens a new field for thoughtful enquiry. Before describing it we must say a few words on *induced* air currents.

1. In 1719 Hawksbee showed that when a current of air was sent through a small box the air within became rarefied. It is a very old lecture-table experiment to suck up and drive a jet of spray out of a bottle by blowing through a tube A directly across the mouth of a tube B dipping into some water. The current of air passing over the open mouth of the tube B carries some of the air from the tube with it, whereby the water rises to B, and is dispersed in a jet of spray.

2. About fifty years ago it was observed at some iron works that a board falling against a blast of air was sucked up to a wall from which the blast issued. The drawing shows a small tube A B threaded through a plate C D, with a card E F in front of it. A current of air blown down the tube will impinge on E F, and be reflected to C D before it passes out. This current will diverge in all directions from B, and will sweep out some air between the card and the plate, causing annular spaces to exist round B as a centre in which the air pressure is diminished. The

FIG. 193.

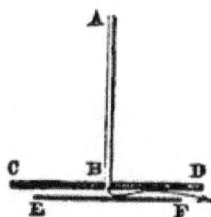

atmospheric pressure outside will therefore support the card. In trying the experiment, some pins must be placed in c D to prevent the card from sliding off laterally.

In both these experiments the original current of air *induced* or set up a current in its neighbourhood, whereby the air around it was set in motion and carried away.

3. The deviation of a musket-ball is due to the fact that an *induced current* is set up by the accidental rotation of the bullet about some axis not coinciding with the line of flight. This rotation is due to accidental friction or impact inside the barrel, and varies with every shot. We will take the worst possible case where the bullet is spinning about a vertical axis and moving in a horizontal line.

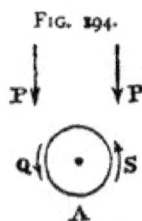

FIG. 194.

The rotation of A sets up a current in the film of air which encircles it, and this current induces a like action in a zone of air of some breadth. As the bullet passes on its course the effect is the same as if we directed a current of air P P upon the bullet encircled with the annular current Q s. On one side of the bullet Q and P move in the same direction, but on the other side s and P are opposed to each other. It is a well-known fact that when two currents of air or gas meet each other they spread out in a lateral direction. That is exemplified in the fish-tail gas-burner, where two currents of gas are thrown obliquely upon each other and spread out into a flat flame. The currents P and s, therefore, cause a lateral deviation in the bullet and would send it towards the left hand in the drawing. The object of the rifling is to cause. the bullet to rotate about an axis coinciding with the line of flight, when the induced current Q s lies wholly in a plane perpendicular to P P, and there can be no lateral deviation from the cause described. The so-called *derivation* of a rifle-bullet is due to a different action.

4. Siemens' *steam jet exhauster* may be taken as the best

of a number of contrivances for setting up induced currents of air by a jet of steam, and was referred to in the introductory chapter.

Here a thin annular jet of steam is employed for setting up the induced current. The air is discharged through the centre of the annulus, and also around the outside, somewhat as it is fed to an argand gas-burner. The rationale of the arrangement is described by Mr. Siemens as consisting in the following particulars :—

1. The air passages are contracted on approaching the jet, whereby the velocity of the entering air is approximated to that of the steam, and there is less loss of efficiency by the formation of eddies.

2. The extent of surface contact between the air and the steam is increased by the annular form of the jet.

3. The combined current of air and steam is discharged into an expanding tube, whereby its velocity is gradually reduced and its momentum utilised by being converted into pressure. (See *Art.* 160.)

As an example of what can be done by this apparatus a trial was made in exhausting air from a vessel containing 225 cubic feet. The jet was $\frac{1}{10}$ sq: inch in area, and the vacuum obtained was 15 inches of mercury after the jet had been in action for 3 minutes, the pressure of the steam being 45 lbs.

We are now in a position to understand the manner in which a jet of high-pressure steam will be competent to suck up and discharge a stream of water from the funnel-shaped opening at E. In the sketch the tube A B E dips into water, and the jet of steam issuing through the opening at E drags the air with it until some water is sucked up to the bend, after which time a mixed jet of water and steam will spurt out at E. Some part of the

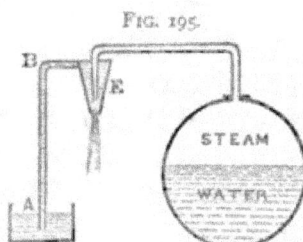

steam will be condensed, and the remainder will break up the water into minute globules which are discharged in the form of spray. If we used air instead of steam and replaced the water by ether we should have a well-known piece of apparatus for sending out a jet of ether spray.

The invention of M. Giffard was the discovery of the fact

FIG. 196.

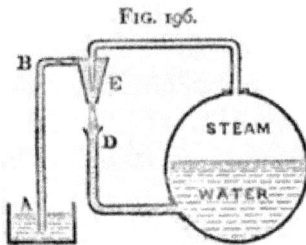

that the mixed jet of steam and water issuing from E was competent to overpower and drive back a simple jet of water issuing from the same boiler in the manner shown in the sketch and that a supply of feed water could be forced into a boiler without any pumping apparatus whatever.

Since action and reaction are equal and opposite, it is abundantly clear that a simple jet of high-pressure steam issuing at E from the boiler could never drive back a jet of water at D issuing from the same boiler under the same pressure. It was a great step in science to discover that the absorption of heat which took place at E when the steam-jet was mixed up with and encumbered by a stream of water could at once furnish a source of energy capable of performing work.

The explanation is to be found in the conversion of the motion of heat at the point E. The steam issuing at E has a velocity many times greater than that of the water forced out at D. This fact will be made clear presently. If the jet of steam could be condensed by an indefinite source of cold it would be converted into a fine liquid line, and the velocity with which its molecules were rushing out would not be changed. The vibratory motion of heat would be taken away, but the onward motion would remain unimpaired. This liquid line would be moving at such a high velocity that it would pierce any jet of water coming towards it from the boiler, very much as if it were a steel wire forcing its way through the mass. We know of no source of cold competent to produce this result, but what really happens is the same in character though less in degree. When the issuing jet of steam is mingled with a fine stream of sufficiently cold water, a portion becomes liquefied, and retains the linear velocity which it had as steam. This higher linear velocity is reduced at once by the sluggish stream flowing up the pipe A B, but, on the whole, the aggregate energy of the water globules flowing onward at E is greater than that of the water jet coming towards them from D. The latter jet is overpowered and driven back, and a quantity of water from the cistern at A is continually driven into the boiler. Referring to the observations on the velocity of efflux of gases in *Art.* 159, let us take steam at an actual pressure of 90 lbs., or 6 atmospheres, which has a volume 321 times greater than water. Here the effective head of water is 5 × (height of water barometer), or 5 × 32 feet, and the velocity of efflux of the water is 8 √ 5 × 32, or 101·2 feet per second. Whereas the velocity of efflux of the steam is 8 √ 321 × 5 × 32, or about 1,813 feet per second. This shows the comparative velocities with which we have to deal.

191. It only remains briefly to describe the apparatus. Steam from a boiler passes through a nozzle F, the area of opening of which is regulated by a rod D, having a conical point. The jet of steam at F propels the feed water through a nozzle at A E, which is opposite a second nozzle B that forms one end of a pipe leading to the boiler. There is a valve at A opening downwards, and the passage from B to A is trumpet-shaped so as to reduce the velocity of the stream before it reaches A.

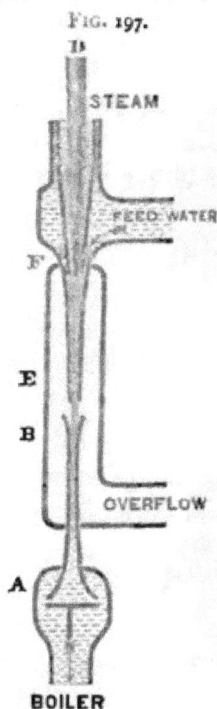

FIG. 197.
STEAM
FEED WATER
OVERFLOW
BOILER

The nozzles E, B are inclosed in a chamber terminating in an overflow pipe by which any surplus water can be got rid of before the apparatus is fairly at work. The amount of steam is regulated by observing the overflow at the nozzle. If there is too much steam the water will not have sufficient energy, and will be forced back into the overflow. If there is too much water the same result will happen, but from a different cause. The whole of the steam will be condensed, and its energy will be dissipated.

The rise in temperature of the feed water shows the amount of energy available for doing work, and it is found that the quantity of water delivered into the boiler increases as the feed water itself is supplied in a colder state. Thus in one case the temperature of the feed water before entering the injector was 60°, 90°, 120°, and the number of gallons of water delivered per hour was 972, 786, 486, respectively. It is a remarkable fact that steam at a low pressure will force water into a boiler against steam at a much higher pressure. Thus steam at 27 lbs. pressure forced water into a boiler where the steam was at 52 lbs. pressure, the temperature of the feed water being raised from 92° to 170° during the operation.

STEAM BREAK FOR LOCOMOTIVES.

192. This is an invention much used on the Continent, where the inclines are severe, and we refer to it because it affords a lesson on the transfer of energy.

If the steam valves were reversed in a locomotive engine, the momentum of the train would, for a time, drive the engine against the steam, and the heated gases would be pumped from the smoke box and forced into the boiler. What would happen is this: the enclosed gases would first of all be compressed and then mixed with steam admitted from the boiler. The gases and steam so mingled would be forced back into the boiler. Thus the energy existing in the moving train might expend itself in compressing and heating the enclosed gases in the cylinders and in further compressing the mingled steam and gas back into the boiler and steam passages. The energy stored up in many tons of moving matter might be converted into its equivalent of molecular motion, with a corresponding disappearance of visible mechanical force. But it would not be practicable to carry out this idea in the manner suggested, for the temperature of the gases in the smoke box is generally about 500° or 600° Fahr., whereby the lubrication of the cylinder and rubbing surfaces would be dried up, and injury would follow. In order to employ reversed working as a steam break for locomotives a jet of water must be first sent into the exhaust pipe. This jet flashes into a fog of mixed steam and water at 212°. It displaces the heated gases and is pumped back into the cylinders. It supplies steam in the place of furnace gas. Passing into the heated cylinder its temperature would be raised, and by compression it would be raised still further. The work thus done in heating the steam and in forcing it to enter the boiler finds its equivalent in the arrested motion of the engine and train, whereby energy is converted into heat and stored up in the boiler, instead of being uselessly expended in grinding the rails and destroying the tyres of the wheels.

The student will now understand that there is no loss of power in reversing the colliery winding engine. The arrested energy is transferred back in another shape into the boiler.

THE MONCRIEFF GUN-CARRIAGE.

193. The principal difficulty attending the use of heavy guns arises from the enormous and destructive force of the recoil. Take the case of the Whitworth 9 inch-gun fired at Shoeburyness in 1868 (*see page* 255). The weight of the gun was 14 tons 8 cwt., while that of the shot was 250 lbs., which is $\frac{1}{129}$ × (14 tons 8 cwt.). At the moment of discharge the energy stored up in the shot was sufficient to carry 250 pounds weight of iron a distance of 11,243 yards, or nearly

$6\frac{1}{2}$ miles. But action and reaction are equal and opposite,
therefore a mass of matter only 129 times as heavy as the
projectile will be suddenly called upon to accept a like quantity of motion.   It is easy to comprehend what a dangerous
force is presented by the recoil under such conditions.

In Captain Moncrieff's carriage the gun is mounted on a
rocking frame A, capable of rolling upon the sides of a horizontal platform C D. A counterweight, sufficient to overcome
the weight of the gun, is so placed that its centre of gravity
lies very nearly in the line joining the points on which the
rocking frame rests.   The trunnions of the gun are directly
over this line.   Two positions of the rocking frame are
shown in the drawing, one before firing, the other during the
recoil.   When the gun is fired, the pressure of the gas
generated by the explosion acts both on the shot and on
the gun; on the shot to propel it outwards, on the gun to
make it recoil.   The gun and rocking frame are thus set in

FIG. 198.

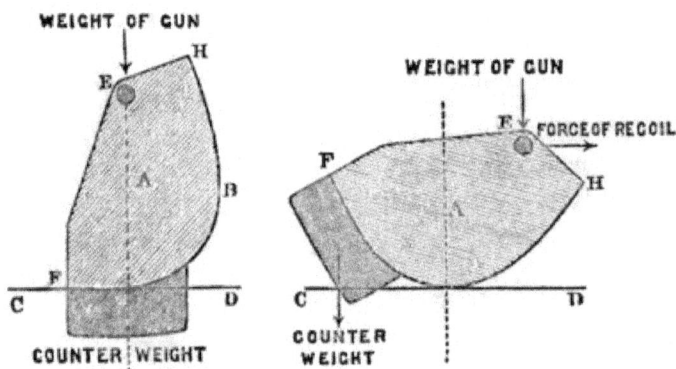

motion. The frame begins to roll very easily until it gets
off the circular part of the frame, which is continued up to
about B, and the curve then changes into a flatter line. The
moment of the counterweight about the point on which the
frame is rolling gradually increases and brings the frame to
rest. When the rolling motion comes to an end the frame

is secured in a position for re-loading. It follows that a gun mounted on one of these carriages may be lifted above the crest of a battery, and will recoil and place itself below the level of exposure after it has been fired. On releasing the frame the counter-weight will restore the gun to the higher position ready for firing.

WATT'S GOVERNOR.

194. The *governor of a steam-engine* was invented by Watt, and has proved of the greatest possible value in steam machinery. It is shown in the drawing, and is a double conical pendulum applied to the regulation of a steam valve. The engine imparts rotation to a pair of heavy balls A, D, swung from the points E F, and connected by the arms K L, M N

FIG. 199.

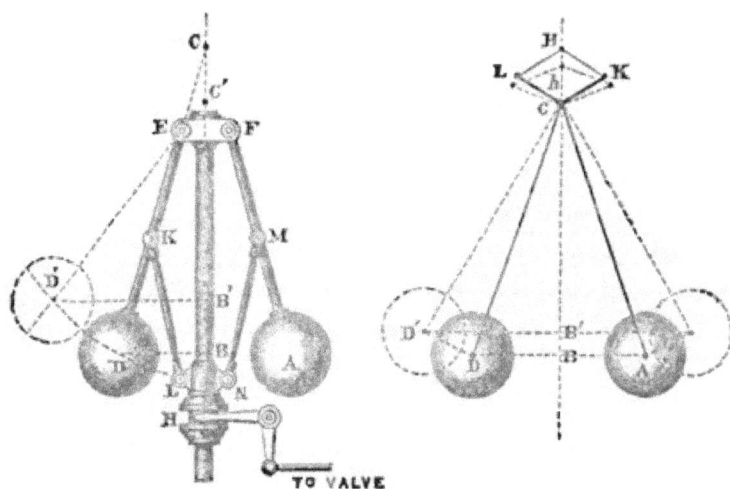

with a sliding collar H, which, in its turn, is connected by levers with the throttle valve regulating the supply of steam to the engine. As the velocity of rotation is increased, the balls fly out, H rises, and the steam valve is partly closed ; whereas, on reducing the velocity of rotation, the balls collapse, and the steam valve is opened more widely.

The most difficult problem which can be presented to the

mechanician is probably that of the exact regulation of revolving mechanism.   In the astronomical clock, or in the
chronometer, we deal out time, it is true, with a precision
which very closely approaches perfection ; but we do so, by
the step by step movement of a train of wheels controlled
either by a pendulum or a vibrating balance.   .

If the governor of an engine were *absolutely perfect*, it
would regulate the velocity of the machine to one uniform
undeviating speed, and would permit of no departure from
that rate of motion.   When the work to be done (technically
called the load on the engine) is varied, the governor would
so adjust the supply of steam as to- keep the machinery
moving at one constant rate.

Watt's governor makes no pretension to realise ideal perfection, and does no more . than *moderate* the inequalities to
which a steam-engine is liable under varying conditions of
load.   It is subject to two principal defects :—

1. Watt's governor cannot prevent a change in the speed
of the engine when a permanent change is made in the
load.   In order to do this the governor should be driven
by some constant force independent of the engine itself,
for it is clear that where a governor is driven by the engine,
and part of the load is taken off, the balls will open more
widely, and can never return to their normal position,
whereby the engine must settle down permanently at a
higher velocity.

2. Watt's governor cannot begin to act until a sensible
change has occurred in the speed; for the balls do not open
without *an increase of velocity*, and the friction of the valve
rod and moveable parts will prevent any motion until the
balls have accumulated an additional store of energy.

In practice, however, this governor is invaluable, not because of the defects pointed out, but in spite of them.   When
any change occurs in the load, the speed of the engine
changes, the balls fly out or collapse, and *moderate* greatly
the amount of such change ; and meanwhile the man in

charge of the engine may be warned by the ringing of a bell and can adjust the supply of steam by hand so as to bring the engine back to the same rate as before.

By comparing the drawing with that in *Art.* 168, the student will thoroughly understand the action of the governor, and he will see that in the ordinary mode of suspension the altitude of the cone changes from c B to c′ B′, when the ball D flies out to D′. Since the time of a revolution varies as the square root of the altitude of the cone, it is bad practice to place the points of suspension E and F at any distance from the vertical spindle. Watt was, of course, aware of this fact, and his original pendulum governor was constructed as in the outline sketch. The balls were swung from the vertical axis, and the valve was connected with H.

SIEMENS' DIFFERENTIAL OR CHRONOMETRIC GOVERNOR.

195. About twenty years ago Mr. Siemens arranged a pendulum governor for steam-engines which was more perfect in its action than the apparatus just described. The pendulum constituting the governor was *driven by a raised weight*, and not by the engine, the result being that the rate of motion of the pendulum could be prescribed definitely beforehand, and would remain invariable. The engine was compelled to adapt its own motion to that of the invariable pendulum, and thus a uniformity of rotation under varying conditions of resistance was arrived at, which could not have been attained by any other known construction. In order to connect the engine and the pendulum, a differential motion was employed,* and the instant that the engine deviated in the slightest degree from the pendulum, the middle wheel in the differential train began to move, and adjusted the throttle valve so as to bring the engine into accord with the pendulum.

The chief peculiarity of the differential governor consists in the fact that the *whole energy* stored up in the revolving

* See the text-book on Mechanism, page 197.

balls is ready to act upon the steam valve at the first instant
that the engine attempts to vary from the pendulum, whereas
in Watt's governor the *additional energy* stored up in the balls
by increased velocity of rotation is the power available to
control the valve. One action is slow and comparatively
feeble, the other is instantaneous and cannot be resisted.

The apparatus now to be described resembles the former
one in everything except the pendulum. Instead of the revolving balls Mr. Siemens employs a cup of parabolic shape
F E, open at both ends, and dipping into water. The cup
rotates about a vertical axis C D, and
drags round some water which assumes the form of a paraboloid, and
soon overflows the rim. The result
is that a stream circulates through
the revolving mass from the base upwards to the edge in the manner
shown in the drawing. In raising
the water to the point of overflow
work is continually being done, and
a resistance is opposed to the driving power which practically remains constant. This resistance may be increased
by directing the overflowing stream against a number of
vanes placed on the outside of the cup, whereby the energy
accumulated in the liquid still further exhausts itself in
opposing the rotation. It is in this respect that the
later governor is superior to that first arranged. The force
which drives the pendulum must necessarily be somewhat
in excess, and the surplus force can only be absorbed by a
resistance set up artificially. In the early form of governor
this resistance was obtained by friction, a doubtful and un-
certain agent ; whereas in the water governor there exists
a resistance capable of absorbing any required excess of
driving force. As a drag to revolving machinery its power
is remarkable. A parabolic cup 4 feet in diameter has been
connected with a tread-wheel, and employed as a regulator

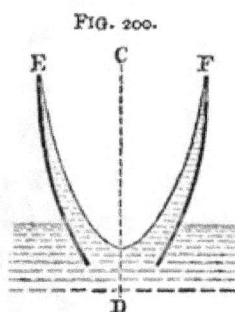

FIG. 200.

of some shafting driven by the wheel. With fifty men at work, the wheel revolved at a certain rate, with 300 men upon it, the rate of revolution did not visibly change, the extra power being automatically absorbed by the hydraulic governor and resulting mainly in the flow of a stronger current of water over the brim.

THE WEIGHTED PENDULUM GOVERNOR.

196. In order to increase the sensitiveness of Watt's governor, it has been proposed to rotate the balls at a high velocity, and to weight them with a load threaded on the central spindle. The general arrangement is shown in the sketch, the balls P, P, weighing from 2 lbs. to 3 lbs. each, and making from 300 to 400 revolutions per minute. The central weight w varies from 50 lbs. to 300 lbs. according to the size of the governor. The balls are con-

FIG. 201.

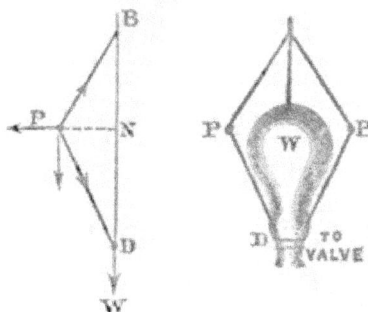

nected to w by 4 equal jointed links, and the friction of the working parts is less than in the ordinary arrangement. The principal advantage gained is an *increase of sensitiveness.* To show this, we refer to the outline diagram and retain the notation of *Art.* 168. Then P is at rest under (1) its weight, (2) the force $\dfrac{P \omega^2 PN}{g}$, (3) T, the tension of B P, (4) T′, the tension of P D. Let $BN = h$, $PBN = \theta$, $PDN = \phi$. Therefore

$$\frac{P \omega^2 PN}{g} = T \sin \theta + T' \sin \phi, \quad P + T' \cos \phi = T \cos \theta,$$

$$2 T' \cos \phi = W, \quad PB \sin \theta = PN = PD \sin \phi.$$

Five equations from which to eliminate θ, ϕ, T, T′, and thereby to obtain P N.

If $\theta = \phi$, we can at once deduce the result $h = \left(1 + \dfrac{\text{w}}{\text{p}}\right)\dfrac{g}{\omega^2}$, whence we infer that the variation of h for any given variation of ω is greater than in the ordinary governor in the proportion of $\left(1 + \dfrac{\text{w}}{\text{p}}\right)$ to 1.

THE CENTRIFUGAL PUMP.

197. We have shown, when treating of *inertia*, that an air pump, capable of ventilating a mine, may be formed by a fan with revolving arms, and we stated that the same principle holds in pumping water. The apparatus then referred to was one form of centrifugal pump.

In order to comprehend the principle here developed, conceive that a ring P is threaded upon a rod D B, pointing towards c and revolving about it. We know that the rod will continually press the ring in a line P Q perpendicular to C D B. The ring will tend always to go forward in a straight line, and will eventually arrive at B. All this has been fully explained.

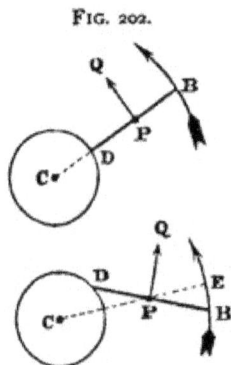

FIG. 202.

Next let the rod D B incline to c P, as shown in the second diagram. The pressure of the rod will be felt in the line P Q inclined to c P at an angle θ.

Let Q be this pressure, then we have a force Q cos θ pushing P directly outwards from the centre. It is evident that P will arrive at a distance c E, which is equal to c B, more rapidly than before. If P were a small portion of water or air, and D B were the blade of a fan the same thing would occur, the inclined arm would act more effectively than the radial arm.

We have now to refer to a simple experiment in the geo-

metry of motion. Let a circular disc be rotated about its
centre c, mark that centre with the
point of a pencil, and then draw
the pencil rapidly from the centre
outwards to the circumference. A
curve will be traced on the disc,
which will take the form c p q if the
disc rotates slowly, or will become
more spiral in character, as shown
by c r s t, when the velocity of
rotation is increased. The pencil
moves radially in a straight line, but it traces a curve by
reason of the increasing linear velocity of each point in the
circle as we pass from the centre outwards. Our conclusion
is that a curved rod will act more effectively than the inclined rod d b, and that if the curvature be regulated to the
velocity of rotation, a sustained and uniform push may be
maintained on each portion of water as it passes from the
centre to the circumference.

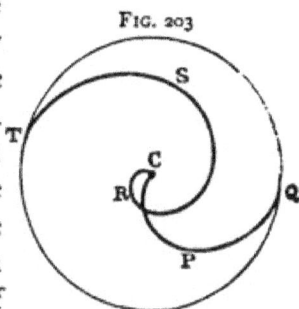

In constructing a pump on this principle we begin with a
circular disc a b, shown in section, which receives the
water pressure on both sides. This is an example of
balanced pressure, and the friction is corre-
spondingly reduced. The disc forms the
central division of a hollow circular box,
the water enters on both sides, as shown
by the arrows, and is forced outwards by
the curved vanes. The three successive
types of vane being given in the diagram
on the next page where the arms are (1)
radial, (2) inclined, (3) curved. In a pump
by Mr. Appold the diameter of the case is 12 inches, that of
the central opening is 6 inches, also there are six arms,
curved backwards and terminating nearly in a tangent to
the circumference of the bounding circle.

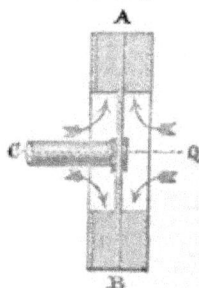

The pump is set into rapid rotation between flat cheeks at

the bottom of a pipe, the circumference of the box being open to the inside of the pipe, and the centre being open to the supply of water about to be lifted. In order that the contrivance may be effective it is necessary to maintain a high linear velocity at the circumference of the disc. We

FIG. 205.

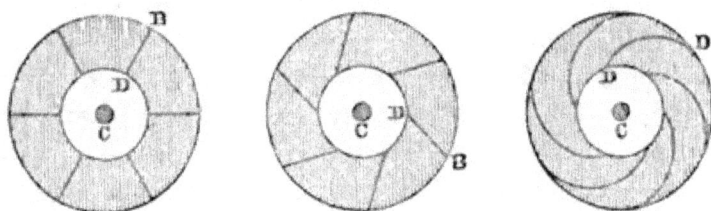

have here an illustration of the conversion of liquid momentum into pressure, and on experimenting with a model we observe the water slowly rising in the stand pipe as the velocity of rotation is increased. It is said that when operating with a pump 12 inches in diameter, the water will stand at 1, 4, 16 . . . feet, with linear velocities of 500, 1,000, 2,000 . . . feet *per minute*, and that an additional linear velocity of 550 feet per minute to each of these numbers will give a discharge of 1,400 gallons per minute at the heights of 1, 4, 16 . . . feet. Since water possesses inertia, any rotation of the mass within the pump itself causes a loss of power, and the object aimed at is to lift the water with as little rotation as possible.

The *Blowing Fan* used in smithies or foundries is substantially a centrifugal pump, but on account of small specific weight of air as compared with that of water the curvature of the arms is not so important.

We have no space to discuss this subject, which presents a wide field for experiment and enquiry.

HUSBAND'S ATMOSPHERIC STAMP.

198. It is a common practice in mechanics to place an elastic spring between the power and the resistance, and thus to avoid the shock and jar consequent on a sudden pull. There are *draw-springs* in railway-trains whereby the pull of the engine is felt first on the spring and then on the carriage. The pull of the chain inside a watch is felt first on a spring concealed within the fusee, and then on the train of wheels. The pull of this spring keeps the watch going while it is being wound up. We have seen that air is elastic in the highest degree, and a mass of enclosed air may furnish an excellent spring which can never wear out.

In crushing tin ore the method in use for centuries has been to lift a series of bars, shod with iron blocks, and to allow these bars or *stamps* to fall by their weight. The stamps are raised by cams placed on a revolving shaft, and the noise and hammering of the cams against the tappets, or projections on the bars, is deafening. The speed which this construction permits is about sixty blows per minute.

In the *atmospheric stamp* an air spring is placed between the driver and the stamp. The stamp is attached to a piston rod Q R which is threaded through a cylinder, having air openings e, f, near the middle, but closed at the two ends. This cylinder is moved up and down by eccentrics on a shaft. As it rises the air in B P becomes compressed, whereby its pressure on the piston overcomes the inertia of the stamp, and the whole rod rises. Soon after the piston has been jerked up by the compressed air, the cylinder will descend; in doing so, it compresses the air in A P and throws the piston down. Thus the stamp is worked by the pressure of the air compressed into each end of the reciprocating cylinder. A fresh supply of air is constantly admitted by the air-holes at e, f. The piston rod is hollow, and serves as a pipe for carrying water necessary for washing away the pul-

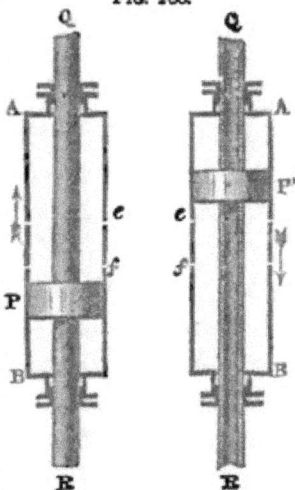

Fig. 206.

verised ore. The heat given out by the air under compression is to a great extent absorbed by the water passing through Q R. This apparatus is so effective that one bar will do the work of ten or twelve ordinary stamps.

THE SMALL-BORE RIFLE.

199. We know very little of the law of resistance of the air to the passage of a rifle bullet, and no mathematician could assign by theory the correct form of a projectile. Nevertheless there is one conclusion which experience confirms, viz., that *the length of a rifled projectile should not be less than 3 times its diameter.* This is the proportion now adopted in the so called *small-bore rifle*, and it may be interesting to the student to learn the manner in which a practical problem of this kind has been worked out by a mechanician.

The task of improving the Enfield rifle was confided to Sir Joseph Whitworth in 1855, and he commenced by building a covered gallery 500 yards long and 20 feet high, wherein to carry on researches without being disturbed by variable currents of air. The rifle bullet was tracked throughout its entire path by light tissue-paper screens, and thus it could be seen whether the point preserved its true direction, or whether it fell over in any degree. The barrel under trial was fixed in a mechanical rest, resembling a a slide-rest, and having true plane surfaces moving on other true planes, whereby the recoil took place in one unchangeable line. To show what can be done with such a rest we may state that it is a common thing to obtain a mean deviation of from 3 to 5 inches with 20 shots from a Whitworth rifle so supported, the range being 500 yards.

In 1855 the Enfield barrel was 33 inches long, the bore was ·577 inch in diameter, the bullet was 1·81 diameters of the bore, and the twist of the rifling was 1 turn in 78 inches. Sir Joseph Whitworth soon satisfied himself that the bullet was too short, and that the twist of the rifling was insufficient. It is to be regretted that we have not space to discuss the question of stability due to rotation, but experience shows that a bullet will certainly fall over in its flight unless the rotation has a definite value which increases, both when the bullet is lengthened, and when its specific gravity is diminished. Thus an increase of $\frac{1}{4}$ inch in the length of the Enfield bullet caused the point to fall over, whereas with an increased twist of rifling as far as 1 turn in 30 inches the lengthened bullet went true to the mark.

In order to ascertain the effect of increasing the twist, barrels were tried having 1 turn in 20, 10, 5 inches, and finally a barrel was rifled to *one turn in one inch*. The bullet fitted the rifle mechanically, and was hardened, or it would have gone out like a musket-ball ; the charge of powder also was greatly reduced. When fired the bullet penetrated 14 half-inch elm planks.

In this way the subject was exhausted, and it was proved in the year 1857 that a rifle-bullet should be *at least* 3 diameters long. This was the form adopted in the Whit-

FIG. 207.

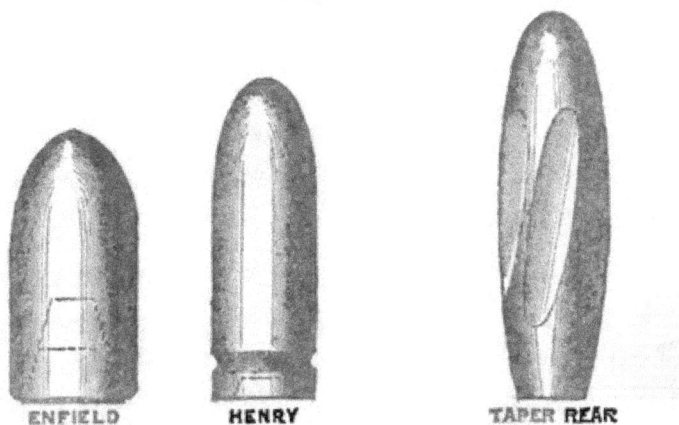

ENFIELD   HENRY   TAPER REAR

worth rifle, the bore being a hexagon with rounded edges, of mean diameter ·47 inch, and the twist being 1 turn in 20 inches. These proportions are now substantially adopted in the best rifles. The drawing shows an Enfield and Henry bullet side-by-side, the diameter of the latter projectile being

X 2

·45 of an inch, and its length 2·93 diameters. In the well-known Metford rifle the bullet is 3·02 diameters in length.

The superiority of the small-bore rifle is proved by the fact that a Whitworth rifle has given an average deviation of $11\frac{1}{2}$ feet at 2,000 yards range, while an Enfield rifle could not touch a target 14 feet square at a distance of 1,400 yards. The penetration of the small-bore is wonderful. In one trial a mechanical-fitting steel bullet in the form of a tube, with a sharp cutting edge, took clean cores out of 34 half-inch elm planks, passing through them just as if it had been a tubular drill.

It will be understood that both the bullets are upset and moulded into the grooves when fired. For heavy guns a projectile must be rifled, and it should have a large amount of bearing surface whereon to receive the twist caused by the rifling. The Whitworth projectile for long ranges is given in the sketch. The bore of the gun is a hexagon with rounded edges, and the rear of the shot tapers in a curve. The range would perhaps be lessened by a mile if the taper form were not adopted. *A heavy piece of ordnance being mechanically the same instrument as a rifle,* the rule as to length applies here equally, and the writer was present at the trial of a 9-pounder Whitworth gun at Southport, in 1872, when a projectile, 4 diameters long, ranged 10,320 yards in the teeth of a strong breeze.

THE METHOD OF PRODUCING A TRUE PLANE.

200. We have spoken so frequently of a plane surface as of a thing well known, that it may be useful, in concluding this treatise, to give an account of the method of producing a *true plane* and to show the connection which exists between *the production of a true plane, and the power of measurement.*

It has been stated by Mr. Maxwell that each particular science advances in the exact proportion in which the power of measurement proceeds, and this is quite as true in mechanics as it is in chemistry, heat, or astronomy. By an effort of genius, and by working in a new and unsuspected direction, Sir Joseph Whitworth has invented a measuring machine which leaves all others at a hopeless distance. It will measure an interval so minute that it cannot even be recognised in the microscope, and since it goes further than the sense of sight will carry us, it does not rely wholly on the eye, but calls in another sense., viz. that of touch, to assist in the observation.

In the year 1840 Sir Joseph Whitworth read a paper at the meeting of the British Association in Glasgow upon ' Plane metallic surfaces and the proper mode of preparing them ;' and at the same time he exhibited specimens of truly plane surfaces.

One of these planes is shown in Fig. 102, and we have pointed out that it rests on three projecting points placed in the angles of a triangle, the object being (1) to secure an equal bearing on each point of support, and (2) to ensure the constant bearing on the *same* points. The plate would otherwise be subject to perpetual variation of form in consequence of the irregular strain occasioned by the change of bearing.

Up to that time plane surfaces had been formed by filing and grinding with emery, a method quite inadequate to produce any good result, and now entirely abandoned. It is only by the process of scraping that a close approximation to a true plane can be obtained. A surface, properly prepared by scraping, will exhibit a vast assemblage of bearing points, evenly distributed, and lying as closely as possible in one true geometrical plane. The mechanic will regard it as a true plane, though it is not an absolutely plane surface, for it is not a perfect reflector, being mottled in every direction by the marks of the scraper. It is a familiar object in the workshop, and the service which it has rendered is incalculable.

The method of obtaining a true plane is described in Mr. Shelley's book, and it is there shown that in order to arrive at *one* true plane, *three* must be operated on simultaneously. It will be understood that *two surfaces can always be rendered identical by scraping.* If surface (1) be coated with a fine film of oil and red ochre, and surface (2) be laid upon it, the prominences on (2) will receive portions of the colouring matter, which can be scraped off, and thus (2) can be worked up to a perfect identity with (1). The process of making a true plane includes three stages, and consists in rendering pairs of surfaces identical according to a uniform method. The surfaces (2) and (3) are brought to coincide with (1), then (2) and (3), which might be both convex or both concave, are compared together and made alike, and finally (1) is made to lie between (2) and (3) in the probable direction of the true plane. The operation is repeated in the same order till the workman is stopped by the inherent imperfection of the material, the goodness of the plane depending on the number of the bearing points, and their distribution at equal distances.

If two well-finished surface plates be wiped with a dry cloth and laid upon one another, the upper plate will appear to float on the film of air between the surfaces, and will move with a touch. If the upper plate be slightly raised and allowed to fall there will be no metallic ring, but the blow will emit a peculiar muffled sound, due to the presence of a cushion of air. If a film of gold-leaf be placed between the plates every atom of it will disappear when the surfaces are rubbed together.

Again, if one surface be carefully slid on the other so as to exclude the air, the plates will adhere together with considerable force by the molecular attraction. The explanation is that the method of scraping has given a vast assemblage of bearing points *evenly distributed*, and lying in one true plane.

TRUE PLANES AT RIGHT ANGLES TO EACH OTHER.

201. The next step is to obtain two true planes at right angles to each other.

For this purpose we operate with three rectangular bars having plane sides, and with two true planes. The bars are placed on a true plane, and another true plane is laid above them. If there be perfect contact between the planes and the bars when the latter are interchanged, partly turned over and reversed, it is at least certain that we have formed bars whose sides are perfectly equal and parallel. But they may not be absolutely rectangular, and in order to ensure this

FIG. 208.

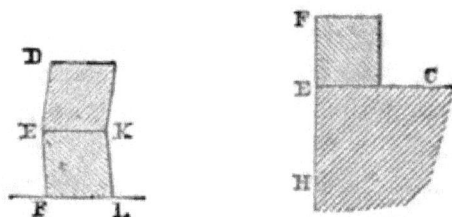

result we can place one bar A on the other B, as in the sketch, and by this process of interchanging, reversing, and testing with true planes, we can finally eliminate any angle, such as D E F, H K L and obtain surfaces which are strictly rectangular and parallel. The true plane on the rectangular finished bar can be copied on the side of a surface plate by causing a true plane to coincide with the surfaces E F, E H, and thus we obtain a surface plate having two plane sides E H, E C accurately at right angles to each other.

In practice there are easier roads to this result, but it is evidently true that the problem can be solved in the manner pointed out.

THE WHITWORTH MEASURING MACHINE.

202. The principle relied upon in the measuring machine is that of employing the sense of touch to aid the sense of sight. It is a matter of observation that if a cylindrical gauge be held in the hand and passed between two parallel and perfectly true planes, these latter surfaces may be so adjusted that it is possible *just to feel* the contact as the cylin-

der moves between them, whereas upon approaching the
planes by $\frac{1}{40,000}$ of an inch, the gauge will pass with greater
difficulty, and the increased pressure necessary to overcome
the resistance is quite apparent.

But it is possible to go much further than this. If we
operate with a light plate of steel whose sides are parallel
true planes, and move the plate as before, we can detect a
resistance when the planes are advanced by an interval much
less than $\frac{1}{40,000}$ of an inch. This resistance is a matter of estimation, dependent on the delicacy of the sense of feeling,
and accordingly Sir Joseph Whitworth determined to test the
tightness of the hold by observing whether or not it was
competent to support a light steel bar with parallel plane
sides, and to hold it suspended between the pressing surfaces. Here is an indication which is quite independent of
the judgment, and which cannot vary. The steel testing
plate has been called by the inventor a *feeling piece*, and is a
light rectangular piece of steel about $\frac{2}{10}$ of an inch thick, $\frac{3}{4}$
of an inch long, with prolonged slender arms. It would be
placed between two planes with one arm resting on a table
and the other arm on the finger of the operator. Its weight
is therefore partly but not wholly supported by the table.
The true planes are now advanced by a very slow movement
until their pressure on the feeling piece is just sufficient to
overcome the force of gravity which comes into play when
the finger is removed. Repeated trials have shown that a
movement of $\frac{1}{1,000,000}$ of an inch is sufficient to release the
feeling piece or to cause it to remain suspended. This
minute difference of distance is therefore a measurable
quantity, and the machine is constructed with the intention
of recording it.

. For the description and drawings of the apparatus we
refer to Mr. Shelley's book. It will there be seen that the
true planes used for measuring are small surfaces at the
ends of rectangular bars having parallel plane sides. *The
ends of each bar are accurately perpendicular to the axis of*

the same ; if they were not so, the construction would fail. The advance of the bars is obtained by a worm wheel having 200 teeth, and rotated by an endless screw having 250 divisions on its graduated micrometer head. The worm wheel advances one of the bars by means of a screw having 20 threads to the inch, and it follows that the rotation of the endless or tangent screw through one division of the scale will advance the bar by $\frac{1}{20} \times \frac{1}{200} \times \frac{1}{250}$ of an inch, that is, by one-millionth part of an inch.

As a piece of constructive mechanism, this instrument stands quite alone, and the exquisite delicacy of its measurement has not even been approached by the most powerful microscopes for reading graduations. It depends, as we have endeavoured to make clear, upon the power of obtaining true planes, and of constructing bars whose plane ends shall be exactly perpendicular to one definite line : thus the power of measurement in this last and highest degree follows as a consequence of the invention of the method of obtaining a true plane surface.

With an account of this apparatus we must conclude our enquiry into the first principles of mechanical science.

LONDON : PRINTED BY
SPOTTISWOODE AND CO., NEW-STREET SQUARE
AND PARLIAMENT STREET